SOLAR INFLUENCES ON GLOBAL CHANGE

Board on Global Change

Commission on Geosciences, Environment, and Resources

National Research Council

NATIONAL ACADEMY PRESS
Washington, D.C. 1994

NOTICE: The project that is the subject of this report was approved by the Governing Board of the National Research Council, whose members are drawn from the councils of the National Academy of Sciences, the National Academy of Engineering, and the Institute of Medicine. The members of the ad hoc group responsible for the report were chosen for their special competences and with regard for appropriate balance.

This report has been reviewed by a group other than the authors according to procedures approved by a Report Review Committee consisting of members of the National Academy of Sciences, the National Academy of Engineering, and the Institute of Medicine.

This work was sponsored by the National Science Foundation, National Aeronautics and Space Administration, National Oceanic and Atmospheric Administration, United States Geological Survey, United States Department of Agriculture, Office of Naval Research, and Department of Energy under Contract No. OCE 9313563.

Library of Congress Catalog Card No. 94-67788
International Standard Book Number 0-309-05148-7

Additional copies of this report are available from:

National Academy Press
2101 Constitution Avenue, N.W.
Box 285
Washington, DC 20055
800-624-6242
202-334-3313 (in the Washington Metropolitan Area)

B-468

Cover artist, Marilyn Marshall Kirkman finds her artistic inspiration in the Rocky Mountain West. Born in Wyoming and now living in Colorado, she is surrounded by a world that demands expression. Vivid color and strong values, eliciting light, convey her message.

Marilyn Kirkman attended Mills College and is a graduate of the University of Wyoming. Having been a teacher and parent, she is now a freelance artist. Largely self-taught, she specializes in watercolor painting. Her work is represented by the Arati Artists Gallery, Colorado Springs, CO.

Printed in the United States of America

Acknowledgments

The Board is deeply indebted to the following group of scientists for their contributions to this report:

JUDITH LEAN, Naval Research Laboratory, *Group Chair*
DANIEL BAKER, University of Colorado
MARVIN GELLER, SUNY at Stony Brook
THOMAS POTEMRA, The Johns Hopkins University
GEORGE REID, National Oceanic and Atmospheric Administration
DAVID RIND, National Aeronautics and Space Administration
RAYMOND ROBLE, National Center for Atmospheric Research
ORAN WHITE, National Center for Atmospheric Research
DONALD WILLIAMS, The Johns Hopkins University
RICHARD WILLSON, Jet Propulsion Laboratory
GEORGE WITHBROE, National Aeronautics and Space Administration
DONALD WUEBBLES, Lawrence Livermore National Laboratory

The National Academy of Sciences is a private, nonprofit, self-perpetuating society of distinguished scholars engaged in scientific and engineering research, dedicated to the furtherance of science and technology and to their use for the general welfare. Upon the authority of the charter granted to it by the Congress in 1863, the Academy has a mandate that requires it to advise the federal government on scientific and technical matters. Dr. Bruce Alberts is president of the National Academy of Sciences.

The National Academy of Engineering was established in 1964, under the charter of the National Academy of Sciences, as a parallel organization of outstanding engineers. It is autonomous in its administration and in the selection of its members, sharing with the National Academy of Sciences the responsibility for advising the federal government. The National Academy of Engineering also sponsors engineering programs aimed at meeting national needs, encourages education and research, and recognizes the superior achievements of engineers. Dr. Robert M. White is president of the National Academy of Engineering.

The Institute of Medicine was established in 1970 by the National Academy of Sciences to secure the services of eminent members of appropriate professions in the examination of policy matters pertaining to the health of the public. The Institute acts under the responsibility given to the National Academy of Sciences by its congressional charter to be an adviser to the federal government and, upon its own initiative, to identify issues of medical care, research, and education. Dr. Kenneth I. Shine is president of the Institute of Medicine.

The National Research Council was organized by the National Academy of Sciences in 1916 to associate the broad community of science and technology with the Academy's purposes of furthering knowledge and of advising the federal government. Functioning in accordance with general policies determined by the Academy, the Council has become the principal operating agency of both the National Academy of Sciences and the National Academy of Engineering in providing services to the government, the public, and the scientific and engineering communities. The Council is administered jointly by both Academies and the Institute of Medicine. Dr. Bruce Alberts and Dr. Robert M. White are chairman and vice-chairman, respectively, of the National Research Council.

Preface

In a series of reports over the past decade, the National Research Council (NRC) has outlined a broad scientific agenda to advance our understanding of the processes of global change. These studies stimulated and nourished the evolution of international efforts centered on the International Geosphere-Biosphere Program (IGBP) and the World Climate Research Program and in our own country supported the development of the U.S. Global Change Research Program. As these programs move rapidly from concept to implementation, the NRC Board on Global Change (BGC) has continued to assess critically the scientific needs. Is the scientific agenda truly comprehensive? Are the priorities appropriate in terms of needs for understanding, scientific opportunities, and technological possibilities? Are there gaps that should be and could be filled? Can recommendations be usefully sharpened and focused?

To address such questions, our Board organized extended ad hoc consultations in a few selected problem areas with informal groups of experts from the scientific community. We focused on problems that were fundamentally important to the program's goals, but were not yet being effectively addressed within the program. Solar influences on the Earth system clearly constituted one such issue. The Sun's energy makes life on this planet possible. Interactions between solar energy and the radiative properties of the atmosphere maintain an equable climate through the greenhouse effect, and there is much concern about human-induced changes in the atmosphere. But the Sun itself is known to vary significantly in its activity. What are the implications of these changes for an already changing planet?

These considerations led the planners of the USGCRP to include solar influences as a major element of the science program. However, the research agenda was at the time relatively undeveloped. What specific research initiatives could be proposed to fill this gap and to improve understanding of the role of the Sun in global change?

In 1990, our Board requested the assistance of a talented group of active research scientists led by Dr. Judith Lean of the Naval Research Laboratory to address these issues. Her group was asked to assist in developing a brief report identifying those aspects of research on the Sun and its interactions with our planet that would contribute to an understanding of global change, together with scientific approaches to developing research plans. It was hoped that these ideas would be useful to the federal agencies as they formulated plans for the U.S. Global Change Research Program.

We are very grateful to Dr. Lean and her collaborators for working with us to develop a set of specific foci for research in this central problem area. We also wish to thank the following individuals:

Dr. Jack Eddy provided the inspiration for this report. The indication that the Sun may be important for the Earth is his vision, carried through the past two decades. His initial concepts, carefully documented in the 1982 Academy Report on Solar Variability, Weather and Climate, laid the groundwork for this more recent assessment of the relationship.

Donald Williams and members of NRC's Committee on Solar Terrestrial Research (CSTR), which together with the Board on Global Change sponsored the ad hoc Group on Solar Influences on Global Change, provided useful critical comments on a draft of the report.

Valuable comments were also provided by many members of the solar and terrestrial research communities, including Linwood Callis, Gizella Dreschoff, Rolando Garcia, John Harvey, James Hecht, Thomas Holzer, Lon Hood, Charles Jackman, Robert Meier, Brian Tinsley, and Edward Zeller. Gary Rottman provided preliminary SOLSTICE data.

We also appreciate the work of Dr. John S. Perry, Mr. Donald Hunt, and Ms. Claudette Baylor-Fleming of the NRC staff in supporting this effort.

Edward A. Frieman, *Chairman*
Board on Global Change

Contents

Solar Influences On Global Change

Executive Summary

Is the Sun an agent of global change? Whether variable energy inputs from the Sun have anything to do with the Earth's weather and climate has been debated contentiously for more than a century. In 1982 a National Academy of Sciences Panel on Solar Variability, Weather and Climate studied the issue in detail, concluding that ***it is conceivable that solar variability plays a role in altering weather and climate at some yet unspecified level of significance.*** In the decade since, monitoring of the Sun and the Earth has yielded new knowledge essential to this debate. There is now no doubt that the total radiative energy from the Sun that heats the Earth's surface changes over decadal time scales as a consequence of solar activity. Observations indicate as well that changes in ultraviolet radiation and energetic particles from the Sun, also connected with solar activity, modulate the layer of ozone that protects the biosphere from solar ultraviolet radiation. This report reassesses solar influences on global change in the light of this new knowledge of solar and atmospheric variability. Moreover, the report considers climate change to be encompassed within the broader concept of global change; thus the biosphere is recognized to be part of a larger, coupled Earth system.

Implementing a program to continuously monitor solar irradiance over the next several decades will provide the opportunity to estimate solar influences on global change, assuming continued maintenance of observations of climate and other potential forcing mechanisms (e.g., greenhouse gases, aerosols, clouds, ozone). In the lower atmosphere, an increase in solar radiation, like the greenhouse gas increase, is expected to cause global warming. In the stratosphere, however, the two effects produce temperature changes of opposite sign. A monitoring program that would augment long term observations of tropospheric parameters with similar observations of stratospheric parameters could separate these diverse climate perturbations and perhaps isolate a greenhouse footprint of climate change. Monitoring global change in the troposphere is a key element of all facets of the United States Global Change Research Program (USGCRP), not just of the study of solar influences on global change. The need for monitoring the stratosphere is also important for global change research in its own right because of the stratospheric ozone layer.

There are no firm plans at present to implement the primary recommendation of this report, a program of continuous monitoring of solar irradiance to provide the data needed to diagnose and interpret solar influences on climate change. Because current solar radiometric techniques are insufficiently accurate, ensuring data continuity over many decades will require a series of space based observations with sufficient temporal overlap for calibration transfer and prevention of data loss from instrument failure. This measurement program may well be precluded by the dearth of access to space.

SCIENTIFIC CONCLUSIONS

Q: Do solar variations directly force global surface temperature?
A: Yes.

Inexorable change is predicted for the biosphere, that sphere of the terrestrial global environment where life exists. It is imperative to reliably detect, understand, and predict climate change arising from increasing greenhouse gases and aerosols in the Earth's atmosphere. This requires that natural climate forcing, particularly solar variability, also be detected and understood. In the study of solar influences on global change, determining the extent to which solar influences modify global surface temperature is a matter of the highest priority.

Energy from the Sun sustains life on Earth. By far the dominant energy input is the visible solar radiation that heats the Earth's land surfaces and the oceans. The atmospheric composition and the distribution of oceans and land masses combine with the solar energy input to determine the radiative balance, and hence the climate, of the Earth's biosphere. A change, ΔS W/m^2, in solar radiation received by the Earth causes climate forcing of $0.7\Delta S/4$ W/m^2 which perturbs directly the equilibrium global temperature by an amount $\Delta T = \lambda^{-1}(0.7\Delta S/4)$°C where the climate sensitivity, λ^{-1}, is currently estimated to be in the range 0.3 to 1.0°C/(W/m^2) (e.g., Wigley and Raper, 1990). While the primary radiative perturbation of ± 3 percent is indisputably the changing insolation generated by the annual cycle of the Earth's elliptical orbit around the Sun, intrinsic changes in the Sun's radiative output also occur on decadal and possibly longer time scales.

The Sun's total radiative input to Earth decreased by about 0.1 percent during 1980-1986 then increased by about the same amount during 1986-1990 (Willson and Hudson, 1991). A 0.1 percent solar irradiance change produces a climate forcing of 0.24 W/m^2. For comparison, the climate forcing by increasing greenhouse gases from 1980 to 1986 was about 0.25 W/m^2 (Hansen et al., 1990). Concomitant increases in atmospheric aerosols may have reduced the net anthropogenic climate forcing to almost half that arising from greenhouse gases alone (Hansen et al., 1993). Thus, during the recent descending phase of the 11-year solar activity, solar forcing canceled much of the net anthropogenic forcing.

The climate system's response to various forcings depends on the history, altitude, and latitude of the forcing and the climate sensitivity, λ^{-1}. While the change in equilibrium global surface temperature associated with a steady climate forcing of 0.24 W/m^2 is estimated to be in the range 0.1° to 0.2°C, the transient response to a periodic 11-year forcing of the same magnitude is assumed to be much less than the equilibrium response because the response time of the climate system is of the order of decades or more (Hansen and Lacis, 1990). However, the true extent to which the climate system's response diminishes or amplifies solar forcing compared with anthropogenic forcings is uncertain.

As records of paleoclimate and historical solar activity have improved, the possibility that variations in solar radiative forcing played a role in past climate change continues to be raised (see, for example, The Royal Society, 1990). There is now clear corroborating evidence from ^{14}C in tree rings and ^{10}Be in ice cores that solar activity during past millennia exhibited a series of minima, each of 40 to 100 years duration, roughly every 200 to 210 years, and that these minima appear to be associated with colder-than-average global temperatures on Earth (Eddy, 1976; Wigley and Kelly, 1990). The coincidence of the Sun's Maunder Minimum with the lowest temperatures of the Little Ice Age is the best documented of such associations in the recent past.

That a physical mechanism might be responsible for the similarity of the historical climate and solar activity records has become more plausible because of observational proof that the Sun's radiative output varied throughout the only 11-year activity cycle during which it has thus far been monitored (Willson and Hudson, 1991). Furthermore, circumstantial evidence suggests that solar irradiance variations may not be limited to the 0.1 percent change detected by contemporary solar monitoring. Observations of Sun-like stars indicate a greater range of activity levels than yet detected in the contemporary Sun (Baliunas and Jastrow, 1990; Lockwood et al., 1992). Also, decreases in irradiance in the range 0.2 percent to 0.3 percent, consistent with the stellar data, are simulated for the Sun's Maunder Minimum by altering the distribution of magnetic features in the solar atmosphere, known to cause much of the 11-year cycle change, within limits defined by independent, spatially resolved solar observations (White et al., 1992; Lean et al., 1992a).

Taken collectively, the above evidence, although circumstantial, does suggest that solar variability could influence future global change, which requires that solar irradiance be properly monitored, understood, and, if possible, predicted. Lack of knowledge of solar influences will limit the certainty with which anthropogenic climate change can be detected. But it is unlikely that solar influences on global change will be comparable to the expected anthropogenic influences. Were solar irradiance to decrease by 0.25 percent over the next 200 years, a value speculated for the Maunder Minimum, the equilibrium global surface temperature is estimated by a general circulation model to decrease 0.46°C (Rind and Overpeck, 1993). This decrease would be too small to offset greenhouse forcing which, by the mid-twenty first century, is expected to have caused a global temperature increase in the range 1.5 to 4.5°C.

Q: Do solar variations modify ozone and the middle atmosphere structure?
A: Yes.

The biosphere, a fragile region in the troposphere where weather and climate are experienced, is protected from solar ultraviolet radiation by a layer of ozone that resides about 30 km above it in the Earth's middle atmosphere. The middle atmosphere is significant for global change not just because of the ozone layer embedded in it; this region is also the upper boundary of the troposphere. Changes in the middle atmosphere, which are known to occur in response to variable solar energy inputs, are suspected of impacting the weather and climate. Determining the extent to which solar variability modifies ozone and the middle atmosphere is therefore the second highest priority in the study of solar influences on global change.

The Earth's middle atmosphere absorbs the Sun's ultraviolet (UV) radiation. Were this radiation able to penetrate to the biosphere, it would damage life on Earth; instead it creates our protective ozone shield. Ozone forms when solar UV radiation (at wavelengths less than 242 nm) dissociates molecular oxygen into oxygen atoms that combine with molecular oxygen to make ozone. Extending outward from the Earth's surface to about 100 km, the ozone layer has its peak concentration at about 30 km. The Sun's UV radiation also creates many of the radical species that subsequently destroy ozone. Most notably, the chlorine (Cl) atom is a product of UV photodissociation of chlorofluorocarbons (CFCs) that have risen to the lower stratosphere following their release near the Earth's surface. Ozone is also destroyed by solar radiation at longer wavelengths and by catalysts produced by energetic particle precipitation. Both the solar ultraviolet radiative energy and the energetic particle output are modulated by solar activity.

The Sun's UV radiation is an order of magnitude more variable than the visible solar radiation that penetrates to the Earth's surface, and these variations generate natural changes in the ozone layer. Specifying natural ozone variability is essential for untangling anthropogenic effects in the long term ozone record. Observational studies (Stolarski et al., 1991; Hood and McCormack, 1992; Randel and Cobb, 1994) signify the response of ozone to solar forcing. From 1986 to 1990 the increase in solar UV radiation in concert with the Sun's 11-year activity cycle is estimated to have increased global total ozone by about 1.8 percent. This approximately offset the suspected anthropogenic decrease of 1.35 percent over the same period (-0.27 percent/year). Other studies suggest that changes in the precipitation of relativistic electrons that penetrate into the middle atmosphere may also play a role in natural ozone variations (Callis et al., 1991).

Current atmospheric loading of chlorine and other anthropogenic radicals is expected to deplete global ozone until the beginning of the twenty-first century. Elimination of CFCs is expected to reverse this downward trend. Determining whether an observed ozone recovery in the twenty-first century is the consequence of successful CFC mitigation or, instead, of increased solar activity will require continuous, reliable monitoring of solar energy inputs to the middle atmosphere. As well as the UV irradiance, sporadic solar influences, such as energetic particles that may destroy ozone for periods of days to many months, must also be understood. This has been clearly demonstrated by a series of high surges of solar activity throughout 1989 (near the peak of the current activity cycle) that may have depleted ozone in the Antarctic (Stephenson and Scourfield, 1991) and at lower latitudes (Reid et al., 1991).

Q: Do solar variability effects in the Earth's upper atmosphere couple to the middle atmosphere and the biosphere?
A: Possibly.

The Earth's lower and middle atmospheres are surrounded by the neutral and ionized medium of the upper atmosphere and its embedded ionosphere, which shelters the biosphere from highly energetic, dramatically varying solar radiation and particles. In the upper atmosphere, temperature, density, and winds are highly responsive to variations in solar energy input. Furthermore, adjacent layers of the Earth's atmospheric envelope are intimately connected.

Highly variable solar inputs to the Earth's upper atmosphere in the form of energetic photons at wavelengths of less than 180 nm and energetic particles cause the global mean exospheric temperature of the thermosphere to vary by about 700 K, from about 600 K during solar cycle minimum to about 1300 K during solar cycle maximum. Physical processes by which this profound solar influence on the Earth's upper atmosphere might impact the biosphere are not established, but radiative, chemical, dynamical, and electrical mechanisms have been identified that couple the upper atmosphere with the middle and lower atmospheres. For example, large variations in both solar high energy photons (X-rays and extreme ultraviolet radiation) and energetic particles initiate significant changes in upper atmosphere odd-nitrogen, which can destroy ozone if transported down to the high-latitude middle atmosphere by the mean circulation pattern (e.g., Huang and Brasseur, 1993). Also, the ionosphere is connected to the troposphere via the global electric circuit.

Solar influences on the upper atmosphere do affect our society in other ways. Activities related to navigation and rescue, defense, and communication rely increasingly on spacecraft technology. Solar energy inputs control the state of the Earth's upper atmosphere where spacecraft orbit, and efficient use of space requires operational understanding of the variability of the upper atmosphere, a part of the global Earth system that is highly sensitive to solar forcing.

In the context of global change *within* the upper atmosphere, knowledge of solar forcing is essential. Since the emission of carbon dioxide infrared radiation is the dominant cooling mechanism throughout the mesosphere and lower thermosphere, releases of trace greenhouse gases from human activity potentially could cause significant changes in the structure of the Earth's upper atmosphere. Whereas the greenhouse effect will cause the troposphere to warm by a few degrees, the global mean thermosphere has been predicted to cool by as much as 50 K in response to projected doublings of carbon dioxide (CO_2) and methane (CH_4) concentrations from present levels (Roble and Dickinson, 1989). Accompanying redistributions of major and minor constituents may decrease satellite drag by up to 40 percent; affect the propagation of atmospheric tides, gravity waves, and planetary waves into the thermosphere from the lower atmosphere; modify the thermospheric circulation; and change the electrodynamic structure. Such anthropogenic effects on the upper atmosphere will be superimposed on large natural variability caused by solar forcing. It is unknown whether these anthropogenic changes could alter the couplings between the upper, middle, and lower layers of the atmosphere. Monitoring the upper atmosphere in the light of its natural variability is therefore important.

Q: Do solar variability effects in the Earth's near-space environment couple to the biosphere?
A: We don't know.

Surrounding the Earth and its atmosphere is the geospace environment of the magnetosphere, composed of solar and space plasmas and energetic particles. Shaped primarily by the Earth's magnetic field and its interaction with the solar wind, the magnetosphere is the primary receptor of the highly variable mass, momentum, and energy from the solar wind. It is tightly coupled to the upper atmosphere and is also involved in the global electric circuit. Any study of solar influences on global change must therefore consider potential coupling to the biosphere.

The relatively self-contained magnetosphere extends from the upper atmosphere to altitudes of about 10 Earth radii on the Sunlit side of the Earth and to more than 1000 Earth radii on the nightside. The global topology of this region is organized by the dipole magnetic field intrinsic to the Earth, which extends far into space and serves to deflect the onrushing plasma, or solar wind, that emanates from the solar corona. The solar wind flows continually over, around, and into the terrestrial magnetosphere, and in so doing continually imparts mass, momentum, and energy to the system. The added energy must then be dissipated either continuously or sporadically. An example of such dissipation is geomagnetic storms, major disturbances in the magnetosphere that manifest themselves by large variations (for periods of hours to days) in the magnetic (and electric) fields surrounding the Earth.

Whether solar forcing of the Earth's near-space environment couples through the upper atmosphere to the biosphere in a way that would be important for global change remains unknown. The extent to which energetic particles couple into the Earth's system depends on the geospace medium through which they must travel. High energy solar protons have been observed to modify ozone concentrations in the middle atmosphere; relativistic electrons precipitating from the magnetosphere may also play a role (Baker et al., 1987). Human activities, such as navigation and resource exploration, can be significantly affected by magnetic field variations associated with geomagnetic storms. In particular, bursts of geomagnetic activity precipitated by solar events can induce current surges that may disable power grids, as was the case in March 1989 (Allen et al., 1989).

Q: Do we need to improve our knowledge of the variable Sun to understand and predict solar influences on global change?
A: Yes.

Solar variability potentially can influence global surface temperatures and middle atmosphere ozone concentrations. The goal of ultimately predicting this influence makes it essential that we learn how and why the Sun varies as it does. Relatively high priority must be given to acquiring knowledge of the origin of the solar energy variations that force global change in the Earth's lower and middle atmospheres.

Our Sun is one of many variable stars in the cosmos. Changes in both its radiative and particle outputs originate in what is actually rather common stellar behavior: a cycle in the emergence of magnetic activity with, in the case of the Sun, a period of 11 years. It is the Sun's magnetic flux that generates the dark sunspots and bright faculae that modulate total solar irradiance, providing radiative forcing of climate change. Extensions of magnetic active regions into higher layers of the Sun's atmosphere are enhanced in shorter wavelength, higher energy radiation. The appearance and disappearance of bright active regions throughout the 11-year activity cycle control the solar radiative output variations that perturb the Earth's ozone layer and also, more dramatically, the upper regions of the Earth's atmosphere. Energetic particles traveling from the Sun to the Earth are guided by lines of the solar and terrestrial magnetic fields, while the solar wind continually transports plasmas and magnetic fields to the Earth's near-space environment.

For studying solar influences on global change, a continuous record of the variable energy input that reaches the Earth from the Sun is essential. The observational record is, however, intermittent and extends over only a few solar cycles. Direct measurements, which must be made from space, are difficult and frequently devalued by instrumental uncertainties. Ideally, our record of the Sun's variable energy input to the Earth should extend over all possible time scales and be predictable into the future. It will ultimately be extended through reliable understanding of how and why the Sun, as a star, varies as it does. This knowledge is needed to unravel possible solar forcing in the paleoclimate record and to assess future solar forcing of global change. It may be obtained through analysis and interpretation of solar images and surrogates that connect the changes in global solar energy output to the fundamental physical parameter changes underlying all solar variability, the magnetic field.

Records of solar activity during the past few thousand years can serve as surrogates for solar energy inputs to Earth, providing the physical connections are adequately understood. Preliminary analysis of these records indicates that extrema of solar radiative output variations may indeed have been larger than the changes during the few recent cycles of activity for which direct measurements exist. Over time scales of stellar evolution, observations of Sun-like stars can help to provide limits on solar variations that might have occurred in the past and may be expected in the future. We cannot presume from our limited monitoring of the contemporary Sun over little more than a decade, and during an epoch of relatively high solar activity, that we have yet sampled the range of variability of which the Sun is capable. But we must nevertheless comprehend this variabity to reliably determine solar influences on global change.

RECOMMENDATIONS

The highest priority and most urgent activity for determining solar influences on global change is to:

1. Monitor the total and spectral solar irradiance from an uninterrupted, overlapping series of spacecraft radiometers employing in-flight sensitivity tracking.

So that the long term value of present solar monitoring is not lost, adequate temporal overlap to permit cross-calibration with future observations is critical. This goal must be achieved in an era of decreasing access to space.

In addition, the following activities will be needed to properly monitor, understand, and predict solar influences on global change.

Pursuit of recommendations 2 to 6 is essential to the interdisciplinary research effort needed to provide an adequate scientific basis for global change policymaking. The actions of recommendations 7 to 12 are essential to ensure a complete understanding of all potential coupling mechanisms.

2. Conduct exploratory modeling and observational studies to understand climate sensitivity to solar forcing.

3. Understand and characterize, through analysis of solar images and surrogates, the sources of solar spectral (and hence total) irradiance variability.

4. Monitor, without interruption, the cycles exhibited by Sun-like stars and understand the implications of these observations for long term solar variability.

5. Monitor globally, over many solar cycles the middle atmosphere's structure, dynamics, and composition, especially ozone and temperature.

6. Understand the radiative, chemical, and dynamical pathways that couple the middle atmosphere to the biosphere, as well as the middle atmosphere processes that affect these pathways.

7. Monitor continuously, with improved accuracy and long term precision, the ultraviolet radiation reaching the Earth's surface.

8. Understand convection, turbulence, oscillations, and magnetic field evolution in the solar plasma so as to develop techniques for assessing solar activity levels in the past and to predict them in the future.

9. Monitor continuously the energetic particle inputs to the Earth's atmosphere.

10. Monitor the solar extreme ultraviolet spectral irradiance (at wavelengths less than 120 nm) for sufficiently long periods to fully assess the long term variations.

11. Monitor globally over long periods the basic structure of the lower thermosphere and upper mesosphere so as to properly define the present structure and its response to solar forcing.

12. Conduct observational and modeling studies to understand the chemical, dynamical, radiative and electrical coupling of the upper atmosphere to the middle and lower atmospheres.

1

Introduction

THE COUPLED SUN-EARTH SYSTEM

The Earth environment as we know it exists because of the energy it receives from the Sun. Radiant energy from the Sun powers the atmospheric and oceanic circulations that profoundly influence the state of the biosphere. Without solar radiation, photosynthesis would cease. Solar radiation and high energy particles impinge continually on the envelope of gases and plasma that surrounds and protects the narrow habitable layer of the Earth's surface. Changes in the amount of solar energy input to the total Earth system are caused by three main mechanisms: i) geometric factors related to the Earth's inclination and orbit around the Sun (which alter the distribution of radiation incident on the Earth), ii) processes in the Earth system itself (which regulate the amount of energy received by the Earth), and iii) variations in the activity of the Sun (which modulate the energy emitted by the Sun).

Geometric relationships modulate solar inputs to the Earth. The seasonal progression of weather is controlled by the tilt of the Earth's axis of rotation relative to the direction normal to the Earth's orbital plane and by orbital eccentricity and precession. In addition, small periodic variations in the Earth's orbital parameters over time scales of tens of thousands of years (Milankovitch cycles) along with associated feedbacks and possible carbon dioxide changes are believed to cause significant variations in the Earth's climate.

Processes within the Earth system regulate the solar energy inputs through numerous feedback mechanisms that influence the greenhouse warming of the Earth. Some of these feedbacks include variations in cloudiness and ice cover that determine the planetary albedo and hence affect the portion of the incoming solar radiation that is available to the Earth system.

Variations in solar energy related to the activity of the Sun can also generate natural changes in the Earth system: assessing the extent of this latter effect is the topic of this report.

There is no doubt that solar variability alters the energy input to the global Earth system, which is considered here in the broadest sense to extend from the biosphere, where weather and climate are experienced, to the Earth's near-space environment, some 1000 km above. Both the short-wavelength ultraviolet (UV) radiation and the solar wind and energetic particles from the Sun undergo large changes related to the presence of active regions in the solar atmosphere. These changes cause dramatic variability in the Earth's upper atmosphere, ionosphere, and magnetosphere. Only recently have spacecraft observations revealed that small variations (about 0.1 percent) also occur in the total electromagnetic energy radiated by the Sun. These radiative variations are also connected to the presence of active regions in the solar atmosphere (dark sunspots and bright faculae), and they occur on all time scales observed thus far, from minutes to the Sun's 11-year activity cycle.

The spectrum of the radiant energy incident on the top of the Earth's atmosphere and the change in this radiation during the solar activity cycle are shown in Figure 1.1. Some of the Sun's radiant energy is reflected back into space by the Earth's surface, by clouds, and by aerosols; the remaining portion is absorbed by the Earth's surface and within the Earth's atmosphere. Figure 1.2 illustrates the altitude of unit optical depth. This is the mean altitude at which solar spectral energy is reduced by the Earth's atmosphere to roughly 1/e of its value at the top of the atmosphere. This curve is determined by the concentrations of radiatively absorbing gases in the Earth's atmosphere. Figures 1.1 and 1.2 indicate that the more variable, shorter wavelength solar energy is absorbed at higher altitudes in the atmosphere. Radiation at wavelengths shorter than

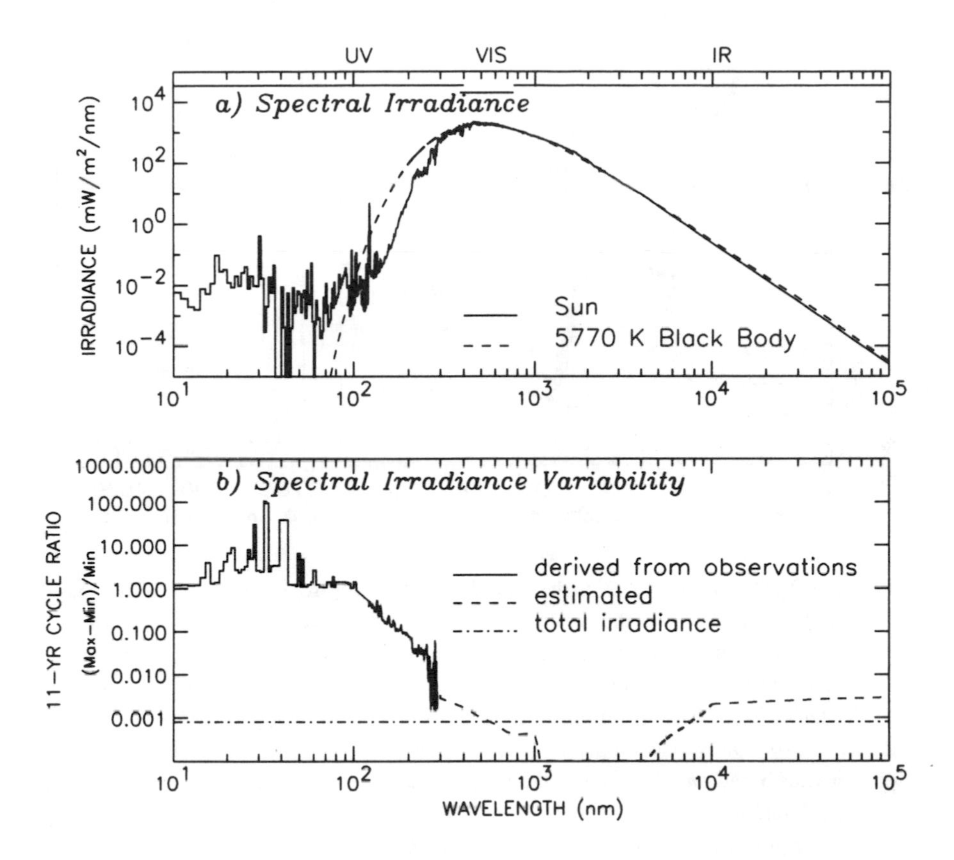

FIGURE 1.1 (a) The Sun's spectral irradiance (solid line, typical of solar minimum conditions) is compared with the spectrum of a black body radiator at 5770 K (dashed line). The broad spectral bands identified along the top of this figure are the ultraviolet (UV), visible (VIS), and infrared (IR). Not shown, at wavelengths longer than the IR, is the microwave or radio portion of the solar spectrum. (b) Approximate amplitude of the Sun's spectral irradiance variation from the maximum to minimum of the 11-year activity cycle. The solar cycle variation in the spectrally integrated, or total, solar irradiance is indicated by the dot-dash line. From J. Lean, Reviews of Geophysics, 29, 506, 1991, copyright by the American Geophysical Union.

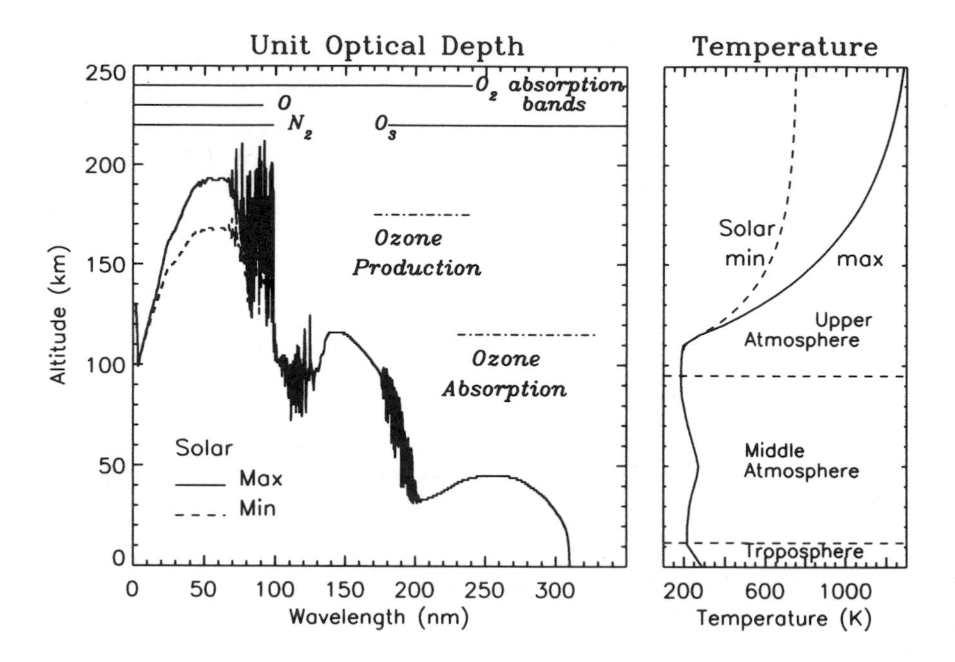

FIGURE 1.2 Shown on the left is the altitude at which the solar irradiance is attenuated by 1/e (unit optical depth) in the Earth's atmosphere, for an overhead Sun. Also indicated are the primary atmospheric absorbing species of the radiation within different spectral bands and the wavelength regions that dominate ozone production and absorption. Adapted from Meier (1991). Shown on the right is the standardized temperature of the Earth's atmosphere from the surface to 250 km. Atmospheric regions, called spheres, are defined by boundaries based on inflections in the temperature profile (at approximately 15 km, 50 km, and 100 km) determined largely by solar radiative heating through gaseous absorption. Reprinted by permission of Kluwer Academic Publishers.

about 160 nm is mostly absorbed above about 100 km (in the thermosphere), where solar variability generates temperature variations of hundreds of degrees. Solar radiation at wavelengths from about 150 to 310 nm is absorbed primarily in the middle atmosphere, which is conventionally defined as being that atmospheric region from about 15 to 100 km, between the troposphere and the thermosphere.

Deposition of the Sun's energetic particle input to the global Earth system is more complicated. Here, interactions with the Earth's magnetic field are important since the charged particles in the solar wind are guided along magnetic field lines. Thus, low latitudes are shielded from much but not all of the incoming charged particles, with most of the energetic particles being guided into the Earth's atmosphere in the polar regions.

Numerous statistical studies have reported variations in atmospheric and hydrospheric parameters that are attributed to solar variability effects on many time scales. Generally speaking, it is much easier physically to relate variations in the Earth's upper atmosphere to known solar activity than is the case for the lower atmosphere and hydrosphere. This is because those energetic inputs from the Sun that show the largest amount of variability (associated with higher energy photons, solar wind, and energetic particles) are usually absorbed in the Earth's upper atmosphere. Solar forcing of the upper atmosphere is thus well recognized and has been verified by the agreement between atmospheric observations and theoretical assessments of the upper atmosphere's response to known solar energy inputs.

Although energetically viable mechanisms for significant solar variability influences on the lower atmosphere and surface of the Earth have yet to be identified, some very interesting associations between solar variability and weather and climate have been suggested. Among these are the cited coincidence between the time of the Maunder Minimum (1645 to 1715) in sunspot activity and the coldest temperatures of the Little Ice Age (Figure 1.3). Another example is the recent work by Labitzke and van Loon (1990, 1993) which suggests an association between solar activity, as measured by the 10.7 cm solar radio flux, and atmospheric temperature changes, with an important role being played by the phase of the quasibiennial oscillation (QBO) in the zonal wind in the tropical lower stratosphere.

The main problem in quantitatively explaining statistical associations between solar variability parameters and sizable climate and weather effects is that the amount of energy in variable solar energy inputs is small compared both to the incoming solar energy itself and to lower atmosphere energetics. Table 1.1 compares the magnitudes of various solar and magnetospheric energy inputs to the Earth system. The total solar radiative energy input per unit area is about 1368 Watts per square meter (W/m^2) in an averaged sense. Observed 11-year cycle variations in total solar irradiance (often referred to as the solar "constant") are about 1.3 W/m^2. This is two orders of magnitude larger than the averaged soft

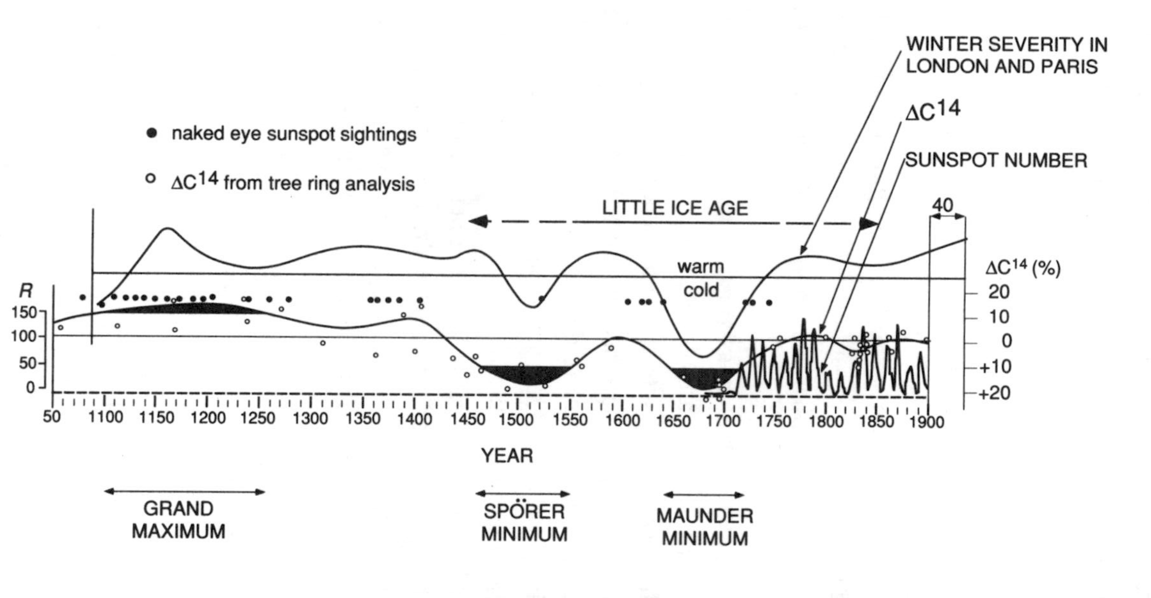

FIGURE 1.3 Relationship between winter severity in Paris and London (top curve) and long-term solar activity variations (bottom curve). The shaded portions of this curve denote the times of the Spörer and Maunder minima in sunspot activity. The dark circles indicate naked-eye sunspot observations. Details of the solar activity variation since 1700 are indicated in the bottom curve by the sunspot number data. The winter severity index has been shifted 40 years to the right to allow for cosmic ray-produced ^{14}C assimilation into tree rings. From J. Eddy, Science, 192, 1189, 1976, copyright by the American Association for the Advancement of Science.

TABLE 1.1 Comparative energy inputs from the Sun to the Earth system and the change in these energy inputs over the 11-year solar cycle. Also indicated are the approximate regions of the Earth system where the energy is deposited.

Source	Energy (W/m^2)	Solar Cycle Change (W/m^2)	Deposition Altitude
Solar Radiation			
total solar irradiance	1368	1.3	surface
UV 200-300 nm	16	0.15	0-50 km
UV 120-200 nm	0.1	0.015	50-120 km
EUV 0-120 nm	0.003	0.005	100-500 km
Particles			
Solar protons	0.002		30-90 km
Proton aurora	0.001 - 0.036		90-130 km
Visual aurora	0.0006 - 0.6		90-130 km
Galactic cosmic rays	0.000007		0-90 km
Joule Heating of Thermosphere			
E= 1 mV/m	0.000014		100-500 km
E=100 mV/m	0.14		100-500 km
Solar Wind	0.0003		above 500 km
Downward Heat Conduction from Magnetosphere	0.00003		above 500 km

X-ray and extreme ultraviolet (EUV) radiation inputs from the Sun and about one order of magnitude less than the solar ultraviolet (UV) radiation that enters into middle atmosphere ozone photochemistry. Solar UV radiation from 200 to 300 nm is believed to vary by a few percent, which implies that the energy associated with changes in this radiation is a factor of 10 or so less than that associated with total irradiance variations. Energetic particle sources possess still less energy.

The direct solar forcing of climate associated with the energy changes in Table 1.1 is currently thought to be smaller than is inferred from some of the observed statistical associations. This makes it difficult to develop viable quantitative models, since more complicated, indirect, amplifying or coupling mechanisms must be invoked. Nevertheless, the more energetic photons and particles that have the largest percentage variations are important candidates for forcing, since they can affect the concentrations of chemical constituents that can possibly redistribute larger amounts of energy. Energy from the Sun, whether as photons, energetic particles, or from solar wind-magnetosphere interactions, and whether deposited at low or high latitudes, is eventually distributed over the entire globe by the continuous motions of the Earth's atmosphere and oceans. Because of this, chemical, radiative, and dynamical perturbations generated by solar variability may be transported to different latitudes and altitudes; this absence of specific spatial and altitude boundaries within the global Earth system means that direct solar forcing of atmospheric regions remote from the biosphere may nevertheless affect it indirectly to some extent.

GLOBAL CHANGE RESEARCH

The goal of the United States Global Change Research Program (USGCRP) is to ***establish the scientific basis for national and international policymaking relating to natural and human-induced changes in the global Earth system*** (Committee on Earth Sciences, 1989). To achieve this goal, the committee defined three specific objectives: i) establish an integrated, comprehensive, long term program of documenting the Earth system on a global scale, ii) conduct a program of focused studies to improve our understanding of the physical, geological, chemical, biological, and social processes that influence Earth system processes and trends on global and regional scales, and iii) develop integrated conceptual and predictive Earth system models.

SOLAR INFLUENCES ON GLOBAL CHANGE: A MAJOR SCIENTIFIC RESEARCH ELEMENT OF THE USGCRP

The need to understand solar variability influences in the study of global change arises because solar-driven global change complicates the detection, understanding, and prediction of anthropogenic forcing. Research efforts toward achieving this understanding might be conceptualized thus:

Monitoring: The solar energy inputs to the Earth system must be measured continuously. Solar phenomena (e.g., sunspots and faculae) that are thought to affect these energy inputs must also be measured. Changes in Earth system parameters must be monitored so that associations can be detected.

Understanding: Once associations between different aspects of solar behavior are established, hypotheses are developed about their physical causes. The predictions of theories based on these hypotheses are then tested against observations in an effort to either prove or disprove the theories. Theories may have to be reformulated in light of new observations. The same intellectual process needs to be followed in formulating theories of the response of the Earth system to variable solar energy inputs. Agreement between theory and observation suggests that a good level of understanding has been achieved.

Predicting: Two types of prediction are possible. One is statistical prediction. In this case, statistical associations are quantified by some sort of regression function which is then used to predict the future. A deeper level of understanding is required to make physical predictions. In this case, quantitative physical laws are formulated and their predictive capability is verified with retrospective data as well as by predictions into the future.

The goal of research in solar influences on global change then is the development of the necessary data bases and understanding of the physical processes that lead to the ability to assess and predict the behavior of the Sun and its influence on the Earth system.

OBJECTIVES OF THE REPORT

This report deliberately focuses first on the most obvious and immediate solar forcing of that part of the Earth's environment where life exists, where understanding solar influences on global change is most important to human welfare and which must thus have high priority. Chapters 2 and 3, therefore, concentrate on solar influences on temperature and composition of the lower layers of the Earth's atmosphere. Chapters 4 and 5 assess solar forcing of higher atmospheric layers and of the Earth's near-space environment and the possible coupling of this forcing to the biosphere. Chapters 4 and 5 do not attempt an exhaustive discussion of all solar-terrestrial connections; this is left, for the most part, to other studies. Chapter 6 discusses knowledge of solar variability itself. Chapter 7 covers strategies for research in solar influences on global change, and recommendations appear in Chapter 8.

The Working Group on Solar Influences on Global Change met twice, in November 1990 and March 1991. Since then the topic has been the focus of three meetings: a Workshop on Solar-Terrestrial Impacts of Global Change, sponsored by the High Altitude Observatory in Boulder, CO, in May 1991; an international symposium on The Sun as a Variable Star: Solar and Stellar Irradiance Variations, International Astronomical Union, Colloquium No. 143, in Boulder in June 1993; and a NATO Advanced Research Workshop on The Solar Engine and its Influence on Terrestrial Atmosphere and Climate, in Paris in October 1993. Proceedings of these three meetings are in preparation. Significant effort has been made to include in this report the relevant results reported at these meetings and in the scientific literature, as of June 1994.

2

Solar Variations and Climate Change

BACKGROUND

As the Sun provides essentially all the energy that drives the Earth's climate system, it is obvious that solar variations have the potential to directly alter climate. Changes in insolation on a variety of time scales have been suggested as causes of known climate change, from the (Milankovitch) orbital cycles of thousands of years (Hays et al., 1976), to the decadal-to-century scale fluctuations typified by the Little Ice Age (Eddy, 1976). During the next 50 to 100 years, the Earth's climate is expected to warm by anywhere from 1.5° to 4.5°C in response to increasing concentrations of greenhouse gases: carbon dioxide (CO_2); methane (CH_4); nitrous oxide (N_2O); chlorofluorocarbons (CFCs) (Intergovernmental Panel on Climate Change, 1992). If solar irradiance were to vary over the next century, natural climate change might also result. Nevertheless, until recently there has been no proof that variations in the Sun's output do in fact occur (as evidenced by the term solar "constant", which is still in widespread use).

Observations of total solar irradiance by spacecraft radiometers (Willson and Hudson, 1991; Hoyt et al., 1992) have now detected decadal variations on the order of 0.1 percent in apparent association with the Sun's 11-year activity cycle (Figure 2.1); larger variations, of the order of a few tenths percent, occur on shorter time scales and are associated with the Sun's 27-day rotation. The magnitude of the 11-year cycle effect

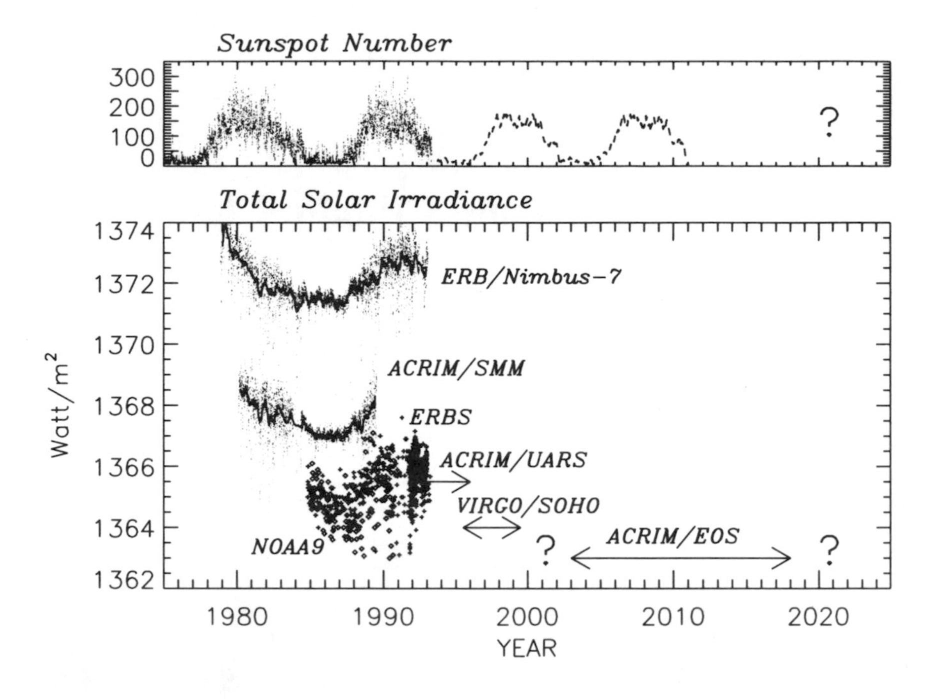

FIGURE 2.1 Contemporary solar activity variations as indicated by the sunspot number (top panel) and changes in total solar radiative output (bottom panel) recorded by the ERB radiometer on the Nimbus 7 satellite, ACRIM I on the SMM satellite and ACRIM II on the UARS, and by the ERBE program (NOAA9 and ERBS). Total solar irradiance is increased during times of maximum solar activity (e.g., 1980 and 1990) and decreased during the intervening minimum. The differences in irradiance levels between the different measurements are of instrumental origin and reflect absolute inaccuracies in the measurements. Proposed future programs to measure total solar irradiance are indicated. Courtesy of J. Lean.

is compared in Figure 2.2 with anthropogenic radiative forcing of climate by increased greenhouse gases and aerosols and by ozone decreases. During the first half of the 1980s, forcing of the climate system by declining solar radiative output was more than sufficient to offset the estimated net anthropogenic forcing.

Despite the similarity of the climate forcings over the decadal time scales shown in Figure 2.2, the magnitude of the climate system's response to solar forcing could be greater or less than its response to anthropogenic forcing. This is because the translation of radiative forcing to surface

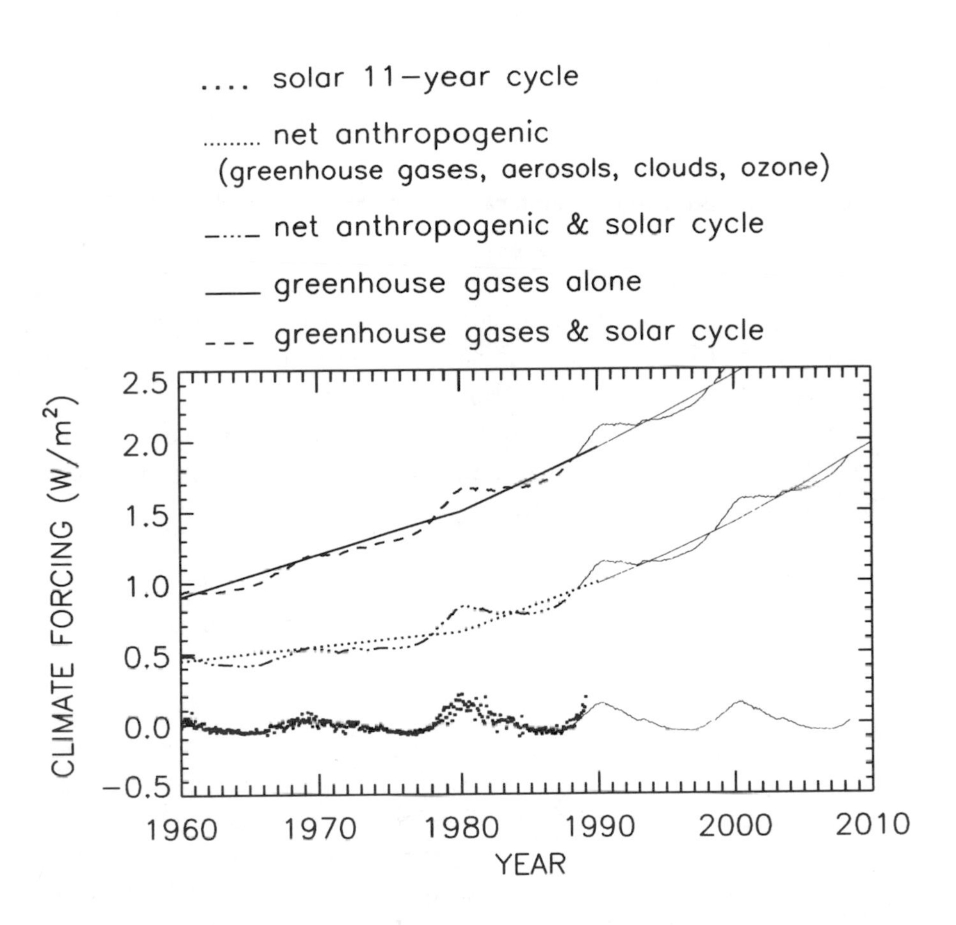

FIGURE 2.2 Estimated climate forcings during the three recent decades of the twentieth-century owing to measured changes in greenhouse gases (solid line), net anthropogenic forcing from greenhouse gases, aerosols, clouds and ozone changes (dotted line), and solar irradiance variations associated with the 11-year solar activity cycle alone (small squares). Combined greenhouse plus solar (dashed line) and net anthropogenic plus solar (dash-dot line) forcings are also shown. In each case, the thin lines are projections. The solar forcing is from the empirical model of Foukal and Lean (1990), which accounts for irradiance changes during the 11-year cycle caused by dark sunspots and bright faculae, but does not include additional variability sources acting on longer time scales. Zero point of solar forcing is the 1978-1989 mean. Adapted from Hansen and Lacis (1990) and Hansen et al. (1993). Reprinted with permission from Nature, Copyright 1990, Macmillan Magazines Limited.

temperature response is highly specific to the altitude, latitude, and history of the forcing (Hansen and Lacis, 1990, Hansen et al., 1993), contrary to the conclusions derived from earlier general circulation model (GCM) comparisons that doubled CO_2 and a 2 percent increase in solar irradiance have an equivalent effect (Hansen et al., 1984). The observed irradiance changes do imply the potential for additional solar forcing in the future, making it incumbent on global change research to monitor, understand, and ultimately predict solar effects on climate. It also makes more compelling the search for a solar signature in the historical climate record.

Understanding solar influences on climate requires the interaction of two primary research areas that are currently quite distinct: the monitoring and assessment of solar irradiance variations, which is reviewed first in this chapter, and the perspective of solar variability and climate from both the paleoclimate record and for future global change, which is discussed subsequently. Origins of the solar radiative output variations are addressed in the broader context of the variable Sun in Chapter 6.

TOTAL SOLAR IRRADIANCE VARIABILITY

Knowledge of the Sun's radiative energy output at all wavelengths is ultimately required for global change research. However, current capabilities for precise determination of this variation exist only at wavelengths shorter than about 250 nm, since at longer wavelengths the measurement uncertainties significantly exceed the amplitudes of the solar variations, which are thought to be less than 1 percent. Measurements of total (spectrally integrated) solar irradiance can be made with two orders of magnitude greater precision and currently provide the primary record of solar radiative output variations.

Contemporary Measurements

During the first three-quarters of the twentieth century, ground based observations were unable to detect total irradiance variations that were unambiguously solar in origin (Frohlich, 1977; Hoyt, 1979; Newkirk, 1983). The two principal limitations were uncertainties due to instrument calibration and to atmospheric interference and attenuation. However, in the recent decade, long term solar monitoring by calibrated experiments flown on spacecraft (to overcome the atmospheric effects) has succeeded in measuring solar irradiance variability on the time scales of the Sun's 11-year activity cycle.

Launched in late 1978, and operational until 1993, the Earth Radiation Budget (ERB) experiment on the Nimbus 7 spacecraft has provided the longest solar irradiance data base (Hickey et al., 1988; Hoyt et al., 1992), although its record (Figure 2.1) is limited by the constraint that the top priority for the Nimbus 7/ERB platform was nadir-looking Earth observations, with only a few minutes per orbit of solar observational opportunity. Coincident with the ERB measurements over most of its lifetime are measurements made by the Active Cavity Radiometer Irradiance Monitor (ACRIM I), launched on the Solar Maximum Mission (SMM) in early 1980 (Willson et al., 1981; Willson, 1984; Willson and Hudson, 1991). This experiment was specifically designed for, and dedicated to, long term, high precision solar total irradiance monitoring; it ceased operation in October 1989 when the SMM spacecraft reentered the Earth's atmosphere.

The Nimbus 7/ERB and ACRIM I results provided the first unequivocal proof of intrinsic total solar irradiance variability, and variations have since been detected on every observable time scale (Figure 2.1). ACRIM's high precision is attributable to its active, electrically self-calibrating cavity (ESCC) solar pyrheliometers and its full-time solar pointing, which provided large numbers of observations. Solar variations measured by ACRIM have been corroborated by the ERB data, with the agreement between the two independent data sets improved by accounting for temperature dependent calibration errors and solar pointing limitations in ERB (Hoyt et al., 1992).

The successor experiments to the Nimbus 7/ERB were the Earth Radiation Budget Satellite (ERBS) and the Earth Radiation Budget Experiment (ERBE) on the National Oceanic and Atmospheric Administration (NOAA)-9 satellite (Lee III, 1990; Lee III et al., 1994). The use of an active cavity ESCC mode for solar observations has improved the quality of the data, but there are operational constraints on solar viewing similar to those with Nimbus 7/ERB, with even less frequent data acquisition opportunities. These latter instruments operate only about every second week and therefore have limited ability to characterize solar rotational modulations, which occur over 27-day time scales. ERBE and

ERBS data (Figure 2.1) both show a decline through the solar minimum period and an increase with the increasing solar activity of solar cycle 22. Differences do exist among the different irradiance data bases in the rate of decrease in cycle 21, the actual occurrence of minimum activity, and the rate of increase in cycle 22. These differences are possibly the result of uncorrected sensor degradation.

In September 1991, ACRIM II was launched on the Upper Atmosphere Research Satellite (UARS). Preliminary data (Figure 2.1) indicate that the UARS/ACRIM II irradiance measurements are systematically lower by about 2 W/m^2 than those of SMM/ACRIM I, whereas Nimbus 7/ERB data indicate solar variations of only a few hundredths of a percent. It would have been preferable to overlap the SMM/ACRIM I and UARS/ACRIM II experiments to provide direct cross-calibration, but the UARS launch delay made this impossible. Thus to preserve the continuity of the ACRIM solar irradiance data base, a third party comparison between ACRIM I and ACRIM II is needed, using the Nimbus 7/ERB or ERBE/ERBS experiments. In the latter case, given the infrequent ERBE/ERBS solar observations, the standard error is estimated to be some 30 times larger than a direct ACRIM I/II comparison.

The data shown in Figure 2.1 indicate that the average solar irradiance declined systematically from 1980 until mid-1986 at a mean rate of 0.015 percent per year. The irradiance minimum in 1986 occurs near the activity cycle minimum of September 1986 (as indicated by the sunspot number data in Figure 2.1). The subsequent rapid increase, corresponding to the buildup of solar activity in solar cycle 22, becomes clearly visible in 1988, continuing to the cycle 22 maximum. Declining values in the latter half of 1992 herald the approach of the next solar activity minimum, expected in 1995-1996. Taken together, the solar radiometer data indicate that the amplitude of the recent 11-year irradiance cycle is about 0.1 percent, disregarding the high Nimbus 7/ERB values in the early years of that record, where the uncertainties are large because of the need to remove significant instrumental effects from the measurements (see Hoyt et al., 1992).

While the ACRIM I, Nimbus 7/ERB, ERBS, and ERBE sensors indeed show similar solar cycle variations of about 0.1 percent (aside from the high Nimbus 7/ERB data in 1978-1979), their absolute solar irradiance values range over some 6 W/m^2, due to absolute calibration uncertainties.

The current inaccuracies of the total solar irradiance measurements, which are typically ± 0.2 percent or larger (Willson, 1984; Luther et al., 1986), are more than twice the downward trend seen from 1980 to 1985. That the uncertainties in measurements made by state-of-the-art solar radiometers are significantly larger than their long term precision, and than the changes caused by solar variability, has important consequences for the continuation of the irradiance data base. In the absence of a third party comparison between ACRIM I and ACRIM II, a decade of solar monitoring would have been terminated, since the solar radiometers lack the accuracies to measure real solar changes smaller than a few tenths percent, twice the 11-year irradiance cycle.

Instruments such as ACRIM and ERB record the variation in the total electromagnetic energy from the Sun without identifying the wavelengths of the radiation at which the variations are occurring. About 99 percent of the total solar irradiance signal is from radiation at wavelengths longer than 300 nm, radiation that penetrates to the troposphere and the Earth's surface. However, shorter wavelength, more variable solar UV radiation (Figure 1.1), which is absorbed primarily above the troposphere (Figure 1.2), contributed approximately 20 percent of the decline in the total solar irradiance from mid-1981 to 1985 (Lean, 1989). It is not known whether the entire solar spectrum varies in phase with solar activity, or how energy might be redistributed within the spectrum. Percentage variations at longer wavelengths are expected to be much smaller than those at UV wavelengths, on the order of a few tenths percent and not necessarily in phase with the activity cycle (Figure 1.1). However, these longer wavelength spectral irradiance variations have yet to be observationally defined.

Implications from Observations of Solar Surrogates

The direct correlation of solar radiative output with solar activity over the 11-year solar cycle is a major discovery from the ACRIM and ERB long term solar monitoring programs. Variations in total solar irradiance occur continuously, on time scales of days to months, in response to episodes of activity throughout the 11-year solar cycle and the modulation of active region emission by the Sun's 27-day rotation. These variations reflect the inhomogeneous emission of radiation on the solar disk. Solar radiation is depleted in active region sunspots and enhanced in active region faculae (Willson et al., 1981; Sofia et al., 1982; Foukal and Lean, 1986; Chapman et al., 1986). From the minimum to the maximum of the 11-year activity cycle there is an increase in active regions, both sunspots and faculae, on the solar disk. Total solar irradiance is thought to be positively correlated with the 11-year solar activity cycle because excess facular brightness, especially from the background active network of bright emission outside of the largest active regions, more than compensates for the sunspot deficit (Foukal and Lean, 1988; Willson and Hudson, 1991). Global perturbations in temperature and/or diameter may also be occurring (Kuhn et al., 1988; Ribes et al., 1989; Kuhn and Libbrecht, 1991; Sofia and Fox, 1994).

To understand the forcing of the climate system by solar irradiance changes, it is necessary to have empirical models capable of extrapolating the radiative output variations to epochs beyond present solar cycles. Knowing that total solar irradiance is enhanced at times of maximum activity, and that these variations appear to arise from the competing effects of two different types of active regions (dark sunspots and bright faculae), suggests that past variations may be reconstructed from historical indicators of solar activity. Empirical parameterizations have been developed to investigate this possibility. The most successful models (Chapter 6) are based on regressions between the ACRIM I or ERB results (corrected for sunspot effects) with specific solar activity indices (derived from the solar He I 1083 nm, Ca II 393.4 nm, and H I 121.6 nm lines) that are considered better surrogates for the total irradiance brightness source than are the classical solar activity indicators, the Zürich sunspot number and the 10.7 cm radio flux (Foukal and Lean, 1988; Livingston et al., 1988).

Many of the major features of the irradiance data have been reproduced by a regression model using the equivalent width (EW) of the solar He I line; these models do not reproduce the high levels of irradiance measured by the radiometers in 1979-1980 near the maximum of solar cycle 21. Also, there are inconsistencies between the Nimbus 7/ERB measurements and model around the time of the cycle 22 activity maximum. Either the empirical relationships differ between solar minimum and solar maximum, and perhaps from one solar cycle to the next, or the irradiance observations are too high because of instrumental artifacts.

When empirical models of total solar irradiance variability developed from the extant spacecraft data are extrapolated over the past century, the long term variations arising from magnetic sunspot and faculae features alone have been no greater than 0.1 percent (Foukal and Lean, 1990). However, as discussed below, limits of solar variability, such as inferred from observations of Sun-like stars, provide circumstantial evidence for a brightness component that has been slowly increasing the total solar irradiance since the Maunder Minimum, a time of reduced solar activity from about 1645 to 1715. With changes in this additional brightness component superimposed on the 11-year cycle variations, a reduction of 0.24 percent is estimated for the Maunder Minimum, relative to the mean of the contemporary 11-year irradiance cycle (Lean et al., 1992a).

Solar observations made by telescopes in the seventeenth century also suggest increased solar diameter and equatorial surface rotation during the Maunder Minimum, compared with the modern Sun (Eddy et al., 1976; Nesme-Ribes et al., 1993). Using apparent solar radius as a surrogate for solar irradiance leads to speculation of a reduction as large as 1 percent during the late seventeenth century (Nesme-Ribes et al., 1993). In addition to uncertainties about the amplitude of solar irradiance values in the Maunder Minimum, there are also differences in reconstructions of the relative temporal variations in the irradiance since then -- over the past 300 years. While derivations based on different solar surrogates -- such as the apparent solar radius record, the length of the sunspot cycle, the sunspot decay rate, or the mean activity level of the 11-year cycle -- do agree about the overall increasing levels of solar activity during the past 300 years, phase differences in specific episodic increases and decreases of activity may be as large as 20 years (Hoyt and Schatten, 1993).

Geophysical Proxies

Relatively continuous, direct records of solar activity exist only since the telescopic discovery of sunspots in the early 1600s. For estimating changes in solar activity over the past several thousand years, other indicators have been proposed, such as variations in cosmogenic ^{14}C in tree rings and ^{10}Be in ice cores (Beer et al., 1988; Suess and Linick, 1990; Beer et al., 1991; Stuiver and Reimer, 1993; Stuiver and Braziunas, 1993). Historical solar activity variations inferred from these cosmogenic isotopes prior to the industrial era are similar (McHargue and Damon, 1991), even though the physical connections between the proxies and solar activity are indirect. For example the ^{14}C record is connected to solar activity as follows. Changes in the solar wind in response to solar activity variations modulate the heliospheric magnetic topology. During times of minimum solar activity, cosmic rays are swept out of the heliosphere less effectively by the solar wind than during maximum solar activity. Thus at solar minima an increased flux of galactic cosmic rays reaches the Earth's atmosphere. This leads to increased production of ^{14}C, which accumulates in the biosphere where it is available for uptake by trees. The similarity between the recent ^{14}C record and the envelope of the sunspot record of solar activity is evident in Figure 1.3. Although many uncertainties exist in interpreting such phenomena, these records offer the potential for gaining improved understanding of solar behavior in the extended past, relevant to global change issues (Wigley and Kelly, 1990; Damon and Sonett, 1991).

Evidence from Observations of Sun-Like Stars

The Sun is a rather common star, and its behavior is thought to be typified by that of stars of similar age, mass, radius, and composition. Routine monitoring of the activity of a selection of Sun-like stars during the past decade has indeed revealed rotational and activity cycles on time scales similar to those seen in the Sun (Radick et al., 1990). Also, observations of Ca II emission in Sun-like stars indicate that 4 out of 13 stars monitored monthly since 1966 exhibited no activity cycle, implying that extended periods of inactivity, as exemplified in the modern solar record by the Maunder Minimum, may be common (Baliunas and Jastrow, 1990). This conjecture is roughly supported by the occurrence of minima that punctuate the ^{14}C geophysical record of solar activity.

In the four Sun-like stars observed to be inactive, Ca II emissions were almost always lower than in the stars that exhibited activity cycles (Baliunas and Jastrow, 1990). White et al. (1992) have shown that the Sun's contemporary Ca II emission corresponds to that of the brighter half of the cycling stars observed by Baliunas and Jastrow (1990) and does not overlap the range of lower Ca II emission typical of noncycling stars. Lean et al. (1992a) investigated the implications of these stellar observations for the Sun's radiative output by utilizing current understanding of the origin of the variations in total solar irradiance and in the Ca II emission from the Sun and stars. Their results, shown in Figure 2.3, suggest that during the Sun's Maunder Minimum the total solar irradiance might have been about 0.24 percent less than its mean value between 1980 and 1990.

Such a decrease is consistent with inferences about the level of solar radiative output during the Maunder Minimum reported by Wigley and Kelly (1990) from the climate record, and also with stellar observations that provide compelling evidence for variabilities of 0.2 percent to 0.5 percent in the luminosity of Sun-like stars (Lockwood and Skiff, 1990; Lockwood et al., 1992). Foukal (1994) notes that the larger luminosity changes observed in Sun-like stars do not necessarily imply equally larger changes in the Sun, at least in the present epoch, since these changes are actually consistent with current understanding of modulation by photospheric magnetism. Also, the variability amplitudes detected in stars likely depend on the observer's viewing angle relative to the stellar spin axis (Schatten, 1993).

SOLAR FORCING OF CLIMATE CHANGE

Variations in solar irradiance may affect the Earth's climate through a direct influence on the global mean temperature or in more subtle ways. The magnitude of climate change that can be associated directly with the changes in total solar irradiance measured during the recent solar activity cycle (about 0.1 percent, see Figure 2.1) is small compared to past climate excursions. Current GCMs estimate that a 2 percent increase in the solar irradiance would produce about 4°C global warming (Hansen et al., 1984). Assuming this result is the right order of magnitude, and that it scales linearly, the 0.1 percent irradiance variation observed by spaceborne radiometers in solar cycle 21 would produce an equilibrium temperature change of about 0.2°C. However, the change from maximum to minimum activity of the 11-year cycle occurs over about five years, too little time to allow for full equilibrium response of the climate system. Furthermore, when averaged over the solar cycle, the effect is reduced by the periodic nature of the forcing, the radiative change during the second half effec-

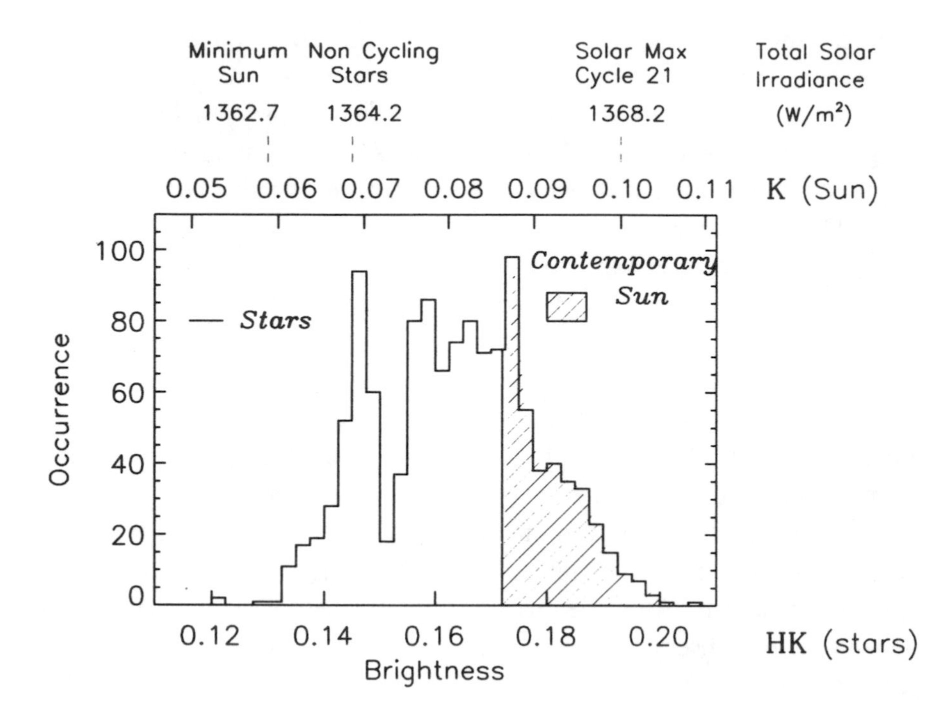

FIGURE 2.3 The distribution of activity in Sun-like stars, determined from observations of Ca II emission (Baliunas and Jastrow, 1990). HK denotes the stellar Ca II emission and K the equivalent solar Ca II emission (see Figure 6.3). Higher values of activity (HK > 0.15) correspond to stars with observed Ca II cycles. Lower values (HK < 0.15) are composed mainly of stars with no periodic variations, and may represent stars sampled in Maunder minima-like phases. The range of activity seen in the present-day Sun (shaded area) is typical of the one-third most active stars in the sample (White et al., 1992). Estimated values of total solar irradiance, from Lean et al. (1992a), were obtained by using the Ca II K data as a proxy for long term brightness variations in the total solar irradiance, and are based on ACRIM radiometry (see Figure 2.1). Courtesy of O.R. White, A. Skumanich, J. Lean, W.C. Livingston and S.L Keil.

tively balancing that during the first half. Thus, direct solar-cycle induced climate change is expected to be less than the variability of ±0.2°C thought to be inherent in the climate system in the absence of external forcing (Hansen and Lebedeff, 1987). To the extent that feedbacks of the climate system are not symmetric, it is possible that solar cycling could produce a net climate forcing that would accumulate over longer times.

From the standpoint of future global climate change, if temperature changes of about 0.2°C were the maximum expected from solar forcing, it might be concluded that solar forcing could be safely ignored, relative to a doubling of greenhouse gases, for which the predicted temperature increase is in the range 1.5°-4.5°C. Yet the climate record suggests that larger effects may have resulted from solar forcing in both the distant and recent past, and that even today unexpected sensitivities may exist (e.g., Wigley and Kelly, 1990; Damon and Sonett, 1991; Labitzke and van Loon, 1993; Thompson et al., 1993). The challenge of expeditious detection of global warming, well in advance of the 50 to 100 year time for greenhouse gas doubling, requires that small (< 1°C) temperature increases be adequately separated into natural and anthropogenic components. Furthermore, as a result of the expected cooling by aerosol and ozone changes, the net anthropogenic climate forcing may be only about half that expected for greenhouse gases alone (Hansen et al., 1993), making the direct solar influence a potentially larger component of the net climate change signal.

Solar-induced changes in the stratosphere could have a variety of indirect influences on the troposphere and climate (e.g., NAS, 1982). Investigations with general circulation models (Kodera, 1991; Rind and Balachandran, 1994) suggest that variations in solar ultaviolet energy input modify the ozone and temperature structure of the stratosphere, affecting the latitudinal temperature gradient. This modifies stratospheric wind speeds and the ability of long-wave energy to propagate out of the troposphere. Altered tropospheric stability affects various tropospheric dynamic processes, including the Hadley cell intensity at low and subtropical latitudes, and low pressure systems in the extratropics.

Paleoclimate, like the recent climate, displays numerous examples of potential interactions between solar radiative forcing and climate. On the longer time scales associated with variations in the Earth's orbital parameters, the insolation gradient between low and high latitudes may be modified, or its seasonal variation changed, which could lead to the growth of ice sheets and altered ocean circulation patterns. Indeed, ubiquitous in climate records are cycles with periods of about 23,000, 41,000, and 100,000 years that are widely attributed to variations in the distribution of insolation with orbital parameters -- the so-called Milankovitch forcing.

For the purposes of the U.S. Global Change Research Program, discussed below are three time scales associated with different mechanisms for solar variability: the relatively recent climate (a hundred to a few thousand years), the weather (tens of years), and orbital variations (many thousands of years).

Solar Irradiance Changes and the Relatively Recent Climate

There has been much speculation that climate changes over the past few thousand years have been the product of variations in the Sun's radiative output. Eddy (1976) pointed out the coincidence in time between the Maunder Minimum of solar activity and the lowest temperatures of the Little Ice Age in Europe and North America (see Figure 1.3). He also presented qualitative evidence that other century-scale variations in climate over the past millennium coincided roughly with variations in solar activity deduced from anomolies in the ^{14}C cosmogenic isotope record. Whereas the long term trend in records of cosmogenic isotopes such as ^{14}C and ^{10}Be reflects, primarily, changes in the Earth's magnetic field that affect the interaction of cosmic rays with the Earth's atmosphere, the wiggles superimposed on the smooth long term trend are believed to occur because of the modulation of the local cosmic ray intensity by magnetic fields embedded in the solar wind, which varies in response to solar activity (Damon and Sonett, 1991; Beer et al., 1991; Stuiver and Braziunas, 1993). Thus, enhanced solar activity corresponds to ^{14}C minima, and the mechanism proposed by Eddy for the apparent relationship between climate and the ^{14}C wiggles involved changes in the total solar irradiance linked to the long term envelope of the 11-year sunspot cycle and reflected in the ^{14}C record.

The extent to which cosmogenic isotope variations really indicate terrestrially relevant variations in solar energy outputs, either radiative or particle, and the scaling of the relationship over long times, is poorly known; the paleoclimate record is similarly somewhat uncertain (Bradley and Jones, 1993). Although results are mixed (Wigley and Kelly, 1990; Crowley and Howard, 1990; Damon and Jirikowic, 1994), there is some suggestion of a relationship; during the past 10,000 years, six of the seven strongest maxima in the ^{14}C wiggles correspond closely to climate minima, reflecting an apparent 200 year cycle in the ^{14}C record that has been recently identified also in the ^{10}Be ice core record.

Reid (1991) has extended the possible relationship between solar variability and climate parameters down to the decadal time scale by pointing out the similarity between the record of globally averaged sea-surface temperature anomalies and the solar cycle envelope over the past 130 years (Figure 2.4). Accounting for the observed sea-surface temperature changes in terms of solar forcing requires a change in solar irradiance of about 0.6 percent. An even more striking relationship between northern-hemisphere surface temperatures and the variation of solar cycle length over the past 130 years (Figure 2.4) has been pointed out by Friis-Christensen and Lassen (1991). The solar cycle envelope and the solar cycle length are related to each other, and both are indicators of the long term variability of solar activity (with longer cycles having smaller amplitudes). The striking correspondence of solar variability indicators with climate parameters provides strong circumstantial evidence for a physical connection, most probably through a variation in solar irradiance, but possibly through an indirect mechanism.

To attribute changes in climate to solar variability, it is first necessary to estimate the magnitudes of changes in both climate and solar radiation and to show that they are consistent. Yet the global temperature changes that have accompanied climatic events during the past millennium are not known in detail. Various figures have been given for the surface temperature excursions of this period, ranging from 0.4-0.6°C (Wigley and Kelly, 1990) to 1°-1.5°C (Crowley and North, 1991). To translate this temperature change into an equivalent solar radiative forcing effect requires that the climate sensitivity be known. Using the value associated with current GCMs of 1°C/(W/m^2) and assuming that the duration of the events was sufficiently long (about 100 years) to allow for a full climatic response, accounting for the entire range of historical climate variability requires changes in total solar irradiance ranging from 0.4 to 1.5 percent. These variations are larger than those observed by the ACRIM I, Nimbus 7/ERB, and ERBS solar radiometers over the past decade (Figure 2.1).

It is not known whether changes in solar irradiance larger than have been observed in the contemporary Sun are plausible on longer time scales, although estimates of the likely variability of 0.24 percent due to changes in sunspots, faculae, and bright magnetic network radiation on century

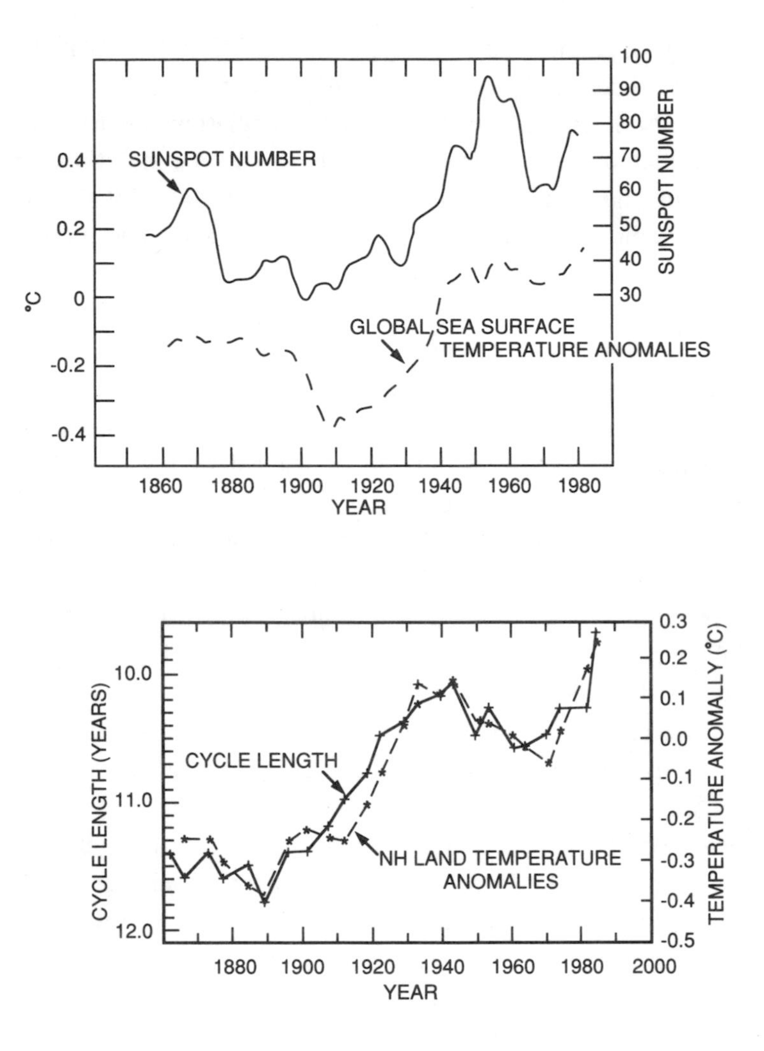

FIGURE 2.4 Solar variability and surface temperature compared in the upper figure are the 11-year running mean of the sunspot number with global average sea-surface temperature anomalies, from G. Reid, J. Geophys. Res., 96, 2835, 1991, copyright by the American Geophysical Union. In a one-dimensional model of the thermal structure of the ocean, consisting of a 100m mixed layer coupled to a deep ocean, and including a thermohaline circulation, a change of 0.6 percent in the total solar irradiance is needed to reproduce the observed variation of 0.4°C in the sea-surface temperature anomalies. Compared in the lower figure are the length of the solar cycle (plus signs) with Northern Hemisphere land temperature anomalies (asterisks), calculated as averages over individual "half" solar cycles (i.e., solar maximum to solar minimum), from E. Friis-Christensen and K. Lassen, Science, 254, 698, 1991, copyright by the American Association for the Advancement of Science.

time scales are somewhat smaller (Lean et al., 1992a and Figure 2.3) than the 0.4 to 1.5 percent needed to explain the paleoclimate record. If the climate sensitivity is greater (one inference from Milankovitch GCM studies; Rind et al., 1989; Phillipps and Held, 1994; discussed below) or the global temperature change smaller than indicated, the required solar variability would be reduced. Furthermore, although GCM climate simulations estimate a mean global temperature reduction of 0.46°C for a solar irradiance reduction of 0.25 percent (Rind and Overpeck, 1993), some regions of the Earth's surface may cool and others warm by as much as 1°C as a result of advective changes caused by differential heating of the land and oceans.

The problem of assessing direct solar radiative forcing of climate change is additionally complicated because the extent to which total solar irradiance variability arises from radiative changes at ultraviolet rather than at visible wavelengths (Lean, 1989) determines the altitude of its direct impact on the global system. If this impact shifts to altitudes mostly above the troposphere, total solar irradiance forcing of surface temperature would be reduced. On the other hand, the amplitude of irradiance variations in the visible and infrared portions of the solar spectrum that directly heat the surface, though thought to be small (e.g., Figure 1.1), is not currently known.

While solar radiative changes are probably not the sole driving force of the historical climate record, they nevertheless will need to be understood and quantified in order to unravel the contribution of solar forcing. Indeed, circumstantial evidence points to a solar forcing contribution to the temperature changes observed over the past century (Kelly and Wigley, 1992; Schlesinger and Ramankutty, 1992) that decreases the predicted temperature change associated with a doubling of atmospheric CO_2 by nearly half (Lacis and Carlson, 1992).

From the perspective of the U.S. Global Change Research Program, it is important to know how solar irradiance variations can be expected to vary in the future and, in particular, the likelihood that events such as another Little Ice Age, will occur in the coming century. Were the only variations in solar radiative output an 11-year cycle with peak-to-peak amplitude of about 0.1 percent, solar forcing could be expected to modulate the net anthropogenic climate forcing as shown in Figure 2.2. But another scenario is that additional solar forcing might arise from longer-term irradiance variations superimposed on the 11-year activity cycle, such as the speculated long term increase in irradiance from the Maunder Minimum to the present Modern Maximum.

Lacking a detailed modeling capability for, and adequate knowledge of, solar processes on which to base predictions, researchers have utilized spectral analysis to develop predictive tools. Phenomena such as sunspot numbers have periodicities on the order of 100, 55, and 11 years, along with the solar magnetic cycle of 22 years (e.g., Berger et al., 1990). Ice core records as well as other climatic data suggest periods of about 80 and 180 years (Johnsen et al., 1970), possibly related to solar activity (Otaola and Zenteno, 1983). Extrapolation into the future of two cycles evident in the ^{14}C record, at 208 years (the Suess cycle) and 88 years (the Gleissberg cycle), suggests that the increasing solar activity that has followed the Maunder Minimum may continue into the early twenty-first century (Damon and Sonett, 1991), with a decline commencing around 2040. But extrapolation of these cycles into the future and prediction of solar effects is a highly questionable procedure, given our lack of knowledge of the fundamental processes involved (see Chapter 6).

Wigley and Kelly (1990) have attempted to assess limits on the role that solar forcing of climate change may play, relative to that of greenhouse gases, during the next 200 years. Analogous to their approach, and consistent with their results, the predictions shown in Figure 2.5 indicate that were the Sun to experience a period of inactivity such as the Maunder Minimum, commencing in the year 2000, and accompanied by reduction in its radiative output of 0.25 percent, the resultant climate forcing would indeed modulate, but not counter, the predicted anthropogenic climate forcing. As noted previously, determining the actual climate impact of the forcings shown in Figure 2.5 (and Figure 2.2) is difficult because of the specific nature expected for the climate system's response to each of the individual forcings.

Solar Activity Cycles and the Weather

There have been many studies of the possible relationships between weather phenomena and the 11-year solar sunspot cycle or the 22-year solar magnetic cycle. Summaries of the results of these studies prior to the early 1980s have been published by Herman and Goldberg (1978) and NAS

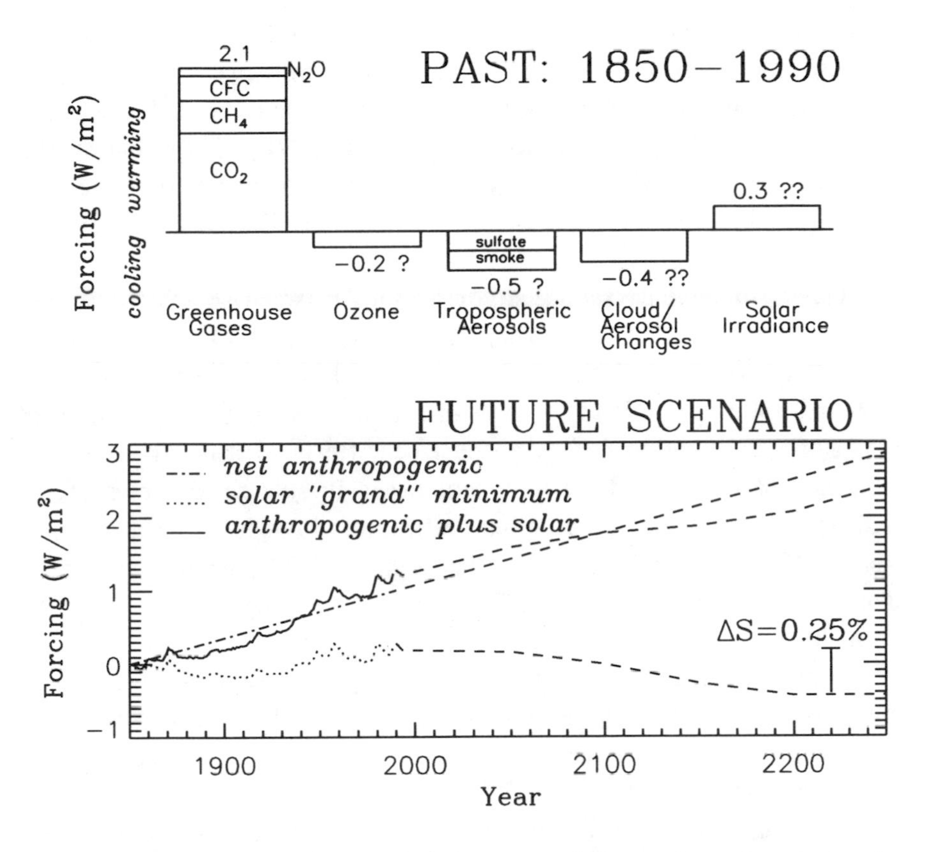

FIGURE 2.5 Climate forcings determined for the past 140 years (upper bar chart) and a scenario for future climate forcing (lower) if anthropogenic forcing continues to increase at its current rate of 1 W/m^2 per 140 years but is partly offset by a solar Maunder Minimum-type event commencing in 2000, taking 200 years to develop. Courtesy of J. Hansen, after Wigley and Kelly (1990).

(1982). While statistical relationships have in some cases been significant, the scientific community as a whole has strongly resisted accepting the findings as proof of a causal relationship, primarily because the mechanisms providing the linkage have not been apparent.

The subject has received new impetus in the past decade, due both to the observation of total and ultraviolet irradiance variations associated with the 11-year solar activity cycle and to observations of a distinct 10-to-12-year oscillation (TTO) in various atmospheric parameters that appear to be in phase with the solar cycle (e.g., Labitzke and van Loon, 1990;

1993) and related to the quasibiennial oscillation (QBO) in tropical stratospheric winds (Figure 2.6). The connection between the TTO and the 11-year solar cycle remains unproven and the statistical validity of the relationship has been debated (Salby and Shea, 1991). Nevertheless, the relationship is considered sufficiently useful to be incorporated in techniques for seasonal forecasting of U.S. weather (Barnston and Livezey, 1989).

Other studies have indicated correlations between solar activity and weather phenomena even when no stratification by QBO phase is made. For example, the mean latitude of winter storm tracks in the north Atlantic appears to shift equatorward at times of maximum solar activity relative to times of activity minima (Tinsley and Deen, 1991). Periods at 11 and/or 22 years have appeared as prominent peaks in spectral analyses of the Earth's surface temperature (Allen and Smith, 1994), sea level pressure (Kelly, 1977), the length of the Atlantic tropical cyclone season (Cohen and Sweester, 1975), ice accumulation data (Holdsworth et al., 1989), drought incidence in the western U.S (Mitchell et al., 1979), the areal extent of North American forest wildfires (Auclair, 1992), global northern hemisphere marine temperatures (Newell et al., 1989), and the separation between annual dust layers in an ice core from the Guliya Ice Cap (Thompson et al., 1993).

However, the basic problem remains: without an understanding of the physical causal connection, the suspicion will persist that the results are the product of a posteriori choices (e.g., Baldwin and Dunkerton, 1989; Salby and Shea, 1991) or are simply the product of natural internal variability (James and James, 1989). Understanding the implied relationships between the Sun and the weather, and the role played by the QBO, would be of enormous benefit, both from the practical standpoint of seasonal forecasting and by enhancing the ability to model and deduce the sensitivity of the climate system to a small external perturbation.

Results of a recent set of GCM studies (Rind and Balachandran, 1994; Balachandran and Rind, 1994) indicate that variations in the middle atmosphere temperature and wind structure associated with the QBO and solar UV irradiance variations *did* impact the troposphere, primarily through alterations in the generation and propagation of the longest tropospheric planetary waves. The resulting longitudinal variations in tropospheric temperature, wind, and geopotential height were similar in

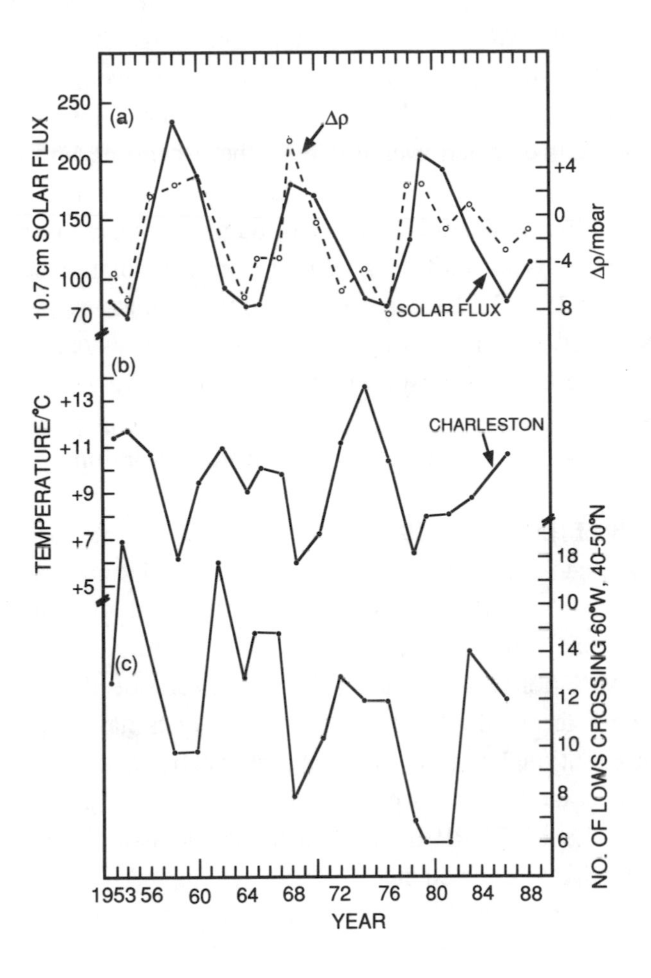

FIGURE 2.6 Compared in a) are the 10.7 cm solar flux and the atmospheric pressure difference [(70°N, 100°W) minus (20°N, 60°W)] in the west years of the equatorial stratosphere quasibiennial oscillation (QBO) in January-February. The changes in the differences between the land (100°W) and sea (60°W) pressures are correlated with the 11-year solar activity cycle. Shown in b) is the surface air temperature at Charleston, South Carolina, during January-February in QBO west years and in c) the number of lows crossing the 60th meridian west between the latitudes of 40°N and 50°N. From Labitzke and van Loon, Phil. Trans, Royal Society London, (1990). Permission granted by the Royal Society of London.

nature and magnitude to those reported by van Loon and Labitzke (1988). The major caveat is the exaggerated UV irradiance variations utilized in the study; nevertheless, knowledge of the response of the troposphere to solar cycle activities that directly affect the middle atmosphere is growing.

In addition, the GCM studies demonstrated that dynamical changes induced by solar cycle variations can affect the radiative properties of the troposphere by influencing cloud and snow cover. Furthermore, the effects do not cancel when averaged over the ascending and descending portions of the cycle. This implies that a time-integrated solar cycle forcing of the climate system is possible through its impact on tropospheric dynamics and feedbacks, rather than through direct insolation perturbation. If so, this would likely have a very different climate impact than the forcing associated with increasing greenhouse gases, whose effect on the middle atmosphere and tropospheric dynamics is entirely different.

Insolation Changes Due to Orbital Variations

The study of Hays et al. (1976) showed that the climate record deduced from deep sea sediments varied with periodicities that generally matched those of the Earth's orbital variations, specifically variations in eccentricity, obliquity (axial tilt), and date of perihelion (Figure 2.7). The theory that orbital variations are indeed the pacemakers of the ice ages has become widely accepted. However, the theory does have problems, both from the observational and the modeling perspectives, that are instructive for evaluating solar influences on global change, and which must be also addressed by the USGCRP in the broader context of the Earth System History USGCRP science element.

The need to understand this issue arises not primarily from the need to predict future climate based on the orbital configurations, but rather from the standpoint of what it implies about the sensitivity of the climate system, and about the ability of climate models to simulate climate sensitivity, since the forcing can be quantified. The last ice age was presumably initiated during the time of strongly reduced summer insolation nearly 110,000 years before the present (BP). The reductions projected for the next 10,000 years are extremely small in comparison and, from this perspective, another ice age is unlikely in that time frame. This is illustrated in Figure 2.7.

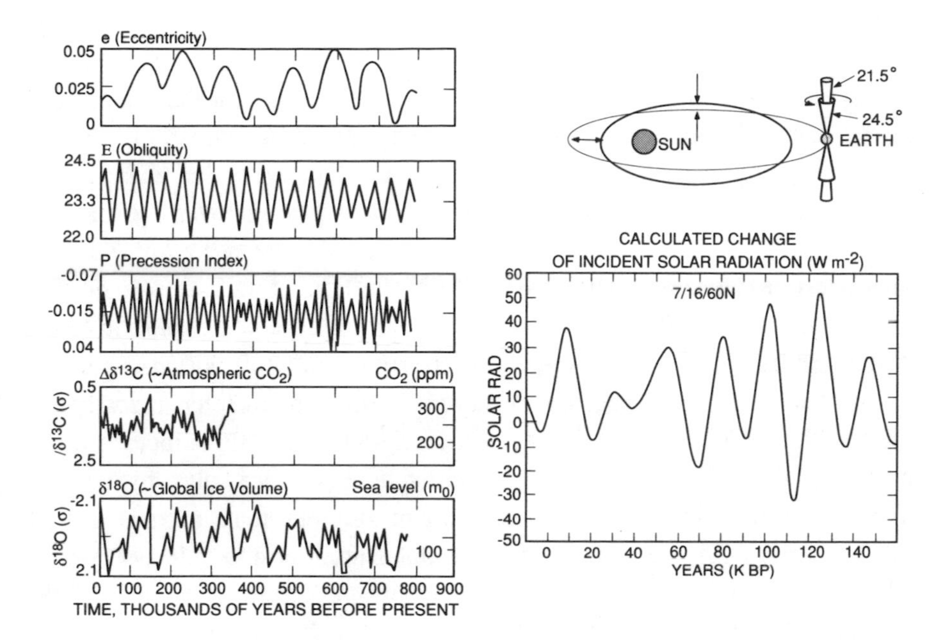

FIGURE 2.7 Orbital (Milankovitch) forcing of climate, as illustrated by the schematic at upper right. Shown on the left are variations in insolation caused by cyclic changes in the Earth's orbital parameters (eccentricity, obliquity, and precession) that are correlated with variations in global ice volume and in atmospheric carbon (from Earth System Science, A Closer View, Report of the Earth System Sciences Committee, NASA Advisory Council, 1988). The calculated changes in northern hemisphere summer solar radiation since 160,000 years BP (from Rind et al., 1989), lower right, indicate extremely small reductions for the next 10,000 years. From this perspective, another ice age is unlikely in that time frame. Courtesy of NASA Advisory Council, 1988, NASA.

Are the changes in insolation effected by the Earth's orbital variations sufficient to have initiated ice ages -- that is, are they the real cause of the glacial/interglacial transitions of the Pleistocene? Melting of the ice sheets occurred 15,000 to 10,000 years BP, coincident with high northern hemisphere summer insolation, in agreement with this hypothesis. However, the southern hemisphere climate also experienced rapid warming in this time interval, when southern hemisphere insolation was at a minimum. The apparent synchronicity of the two hemispheres in their responses to orbital variations, which for the precessional cycle has opposite solar insolation effects in the two hemispheres, has long been a mystery and raises the question of how much of the climate response is actually associated with orbital forcing.

Spectral analysis of the paleoclimate record shows that the maximum power lies in the approximately 100,000 year period, which is of the same order as the Earth's eccentricity variation. However, the changes in eccentricity, on the order of a few tenths of a percent over the past 5 million years, produce little change in net annual solar radiation, so that any possible effects on the seasonal distribution of radiation must be combined with variations in tilt and precession of the Earth's rotation axis, which are larger. Thus it is surprising that the about 100,000 year period dominates in the climate record. Examples of this mismatch can easily be found: the peak of the last ice age, about 20,000 years BP, coincides with a very weak minimum in Northern Hemisphere summer solar insolation, and the deglaciation Northern Hemisphere summer maximum at about 12,000 years BP is no larger than a similar feature at about 30,000 years BP, which did not lead to complete deglaciation. These facts suggest that processes other than direct solar forcing may be responsible for the observed climate record.

Even the timing of the insolation variations relative to the climatic response has been questioned. Winograd et al. (1988, 1992) analyzed the oxygen-18 variations found in a calcitic vein in the southern Great Basin. The uranium series age dates of the calcite vein indicated that major glacial/interglacial transitions occurred some 10,000 to 20,000 years before the solar insolation variations; for example, the peak interglacial in this record appears at 147,000 $\pm$ 3,000 years BP, significantly before the insolation peak. While the relevance of this local record to global temperature and precipitation changes may be in doubt, high sea level stands in the period 135,000 to 140,000 years BP have been found by various researchers (e.g., Moore, 1982). The absolute dating capability associated with the calcite vein is in contrast to the approximate dating techniques associated with the deep sea paleoclimate record, where assumptions about sedimentation rates are fundamental in matching the orbital periodicities.

When the orbital solar insolation variations are incorporated in general circulation climate models, the temperature changes are not sufficient to produce ice sheet growth, especially in regions of low altitude accumulation, as was apparently the case for the Laurentide ice sheet (Rind et al.,

1989; Phillipps and Held, 1994). Either the models are incomplete or orbitally induced solar insolation variations are at best only a catalyst for glacial/interglacial changes. Both of these conclusions have important implications for global change projections: the former implies that contemporary GCMs might not be sufficiently sensitive to solar radiative forcing (whether of orbital or solar activity origin), while the latter emphasizes that it is the climate system feedbacks that are most important in producing climate change, invalidating the use of simple transfer functions between radiative forcing perturbations and climatic responses.

3

Solar Variations, Ozone, and the Middle Atmosphere

BACKGROUND

Weather and climate are experienced in the troposphere, which extends upward to the tropopause at about 15 km. The Earth's middle atmosphere (the stratosphere and the mesosphere) extends from the tropopause to approximately 90 km (Figure 1.2). Understanding the middle atmosphere is crucial for global change studies for two primary reasons. Firstly, this is where approximately 90 percent of the Earth's ozone shield resides and, secondly, the region constitutes the upper boundary of the troposphere. Changes in the ozone layer modulate the amount of ultraviolet (UV) radiation reaching the biosphere and are thus of direct concern for life on Earth. A decrease in column ozone of 1 percent causes the UV dose in the spectral region damaging to deoxyribonucleic acid (DNA) to increase by about 2 percent. Changes in ozone, temperature, and other trace gases have been widely implicated in a variety of the mechanisms by which changes in the middle atmosphere might influence the biosphere by physically coupling to the troposphere and modifying the climate (NAS, 1982; Lacis et al., 1990; Schwarzkopf and Ramaswamy, 1993; Hauglustaine et al., 1994; Rind and Balachandran, 1994). Crucial to these mechanisms is ozone, which is highly responsive to, and also has a controlling influence on, the state of the middle atmosphere. Ozone is directly influenced by changes in solar radiative and energetic particle inputs to this region.

All of the Sun's ultraviolet energy input at wavelengths between about 150 and 300 nm is deposited in the Earth's middle atmosphere. This energy plays an essential role in the chemistry, radiation, and dynamics of the region. Figure 1.2 shows the altitude at which solar radiation (from an overhead Sun) reaches 1/e of its original intensity and gives an approximate measure of the penetration depth for different wavelengths. The inversion in the atmospheric temperature profile at about 15 km (Figure 1.2) that defines the tropopause is a direct consequence of heating by solar UV energy abosorbed by ozone in the middle atmosphere. The only significant solar radiation of shorter wavelength reaching the middle atmosphere is a small and highly sporadic contribution from X-rays, and the strong H I Lyman α line at 121.6 nm.

The ozone layer exists because of the interaction of solar UV radiation with the constituents of the middle atmosphere. Photodissociation of molecular oxygen by solar UV radiation at wavelengths from 170 to 242 nm (in the O_2 Schumann-Runge bands and Herzberg continuum) is the chief source of atomic oxygen and hence ozone production. Formed by combination of atomic and molecular oxygen, ozone is in turn photodissociated, mainly by solar UV radiation at wavelengths between 240 and 300 nm (in the strong O_3 Hartley bands and continuum), but also by longer wavelength visible solar radiation. Ozone's strong absorption in the UV region of the spectrum serves the dual role of heating the middle atmosphere and protecting the surface of the Earth from damaging doses of ultraviolet radiation. Solar UV radiation also creates other important trace constituents, such as chlorine (Cl) and hydroxyl (OH), that participate in catalytic reactions that destroy ozone. Current understanding of the processes affecting ozone and other trace constituents of the middle atmosphere has been reviewed in a number of reports (e.g., WMO, 1988), and is not repeated here.

Stratospheric ozone has received intensive study in recent years, and is the subject of research under the Biogeochemical Dynamics science element of the USGCRP. Ozone is known to be influenced by human related sources such as chlorofluorocarbons (CFCs), carbon dioxide (CO_2), and methane (CH_4) and by natural occurrences such as volcanoes and solar variability. As summarized in Figure 3.1, definite changes in the total amount of stratospheric ozone are thought to have occurred throughout the past few decades. Long term changes in stratospheric temperatures have also been found (WMO, 1988; Randel and Cobb, 1994). A significant fraction of the long term changes in ozone and temperature is thought to be caused by human related emissions (of CFCs and other gases). Recent analysis of data from the Total Ozone Mapping Spectrometer (TOMS) on board the Nimbus 7 satellite indicates a global (65 N to 65 S latitude) ozone decrease of 0.27 percent $\pm$ 0.14 percent per year since 1978 (Stolarski et al., 1991; Hood and McCormack, 1992). This is thought to arise from anthropogenic effects.

A significant fraction of the ozone variance during the past decade is also due to solar forcing (Figure 3.1). The ozone content of the middle atmosphere varies with the 11-year solar cycle because the solar radiation and particle environment responsible for creating and destroying ozone varies. Superimposed on the long term downward ozone trend deduced from TOMS data during the past 11 years is a solar cycle variation whose amplitude is estimated to be 1.8 percent $\pm$ 0.3 percent (Hood and McCormack, 1992). These solar-related changes in ozone exacerbated the downward anthropogenic trend from 1982 to 1986 (the descending phase of the solar activity cycle) and masked it almost completely during 1986-1991 (the ascending phase of the cycle).

Longer-term variations related to solar variability are clearly possible. To develop a reliable understanding of true anthropogenic effects on ozone therefore requires that its natural variability in response to solar influences be fully defined; knowledge of solar-induced changes will be particularly important in the future in verifying whether an observed slowing of the ozone depletion rate is the result of limits on anthropogenic sources or a response to changing solar energy inputs.

As mentioned in Chapter 2, solar-induced variations in stratospheric temperature structure may be affecting the troposphere through modulation of the Hadley circulation in the tropics or planetary wave generation in the extratropics. Changes in the latitudinal temperature gradient in the stratosphere and planetary wave generation in the extratropics affect stratospheric wind speeds and thus the ability of long-wave energy to propagate out of the troposphere, further altering tropospheric dynamical patterns and subsequently radiative parameters (Rind and Balachandran, 1994).

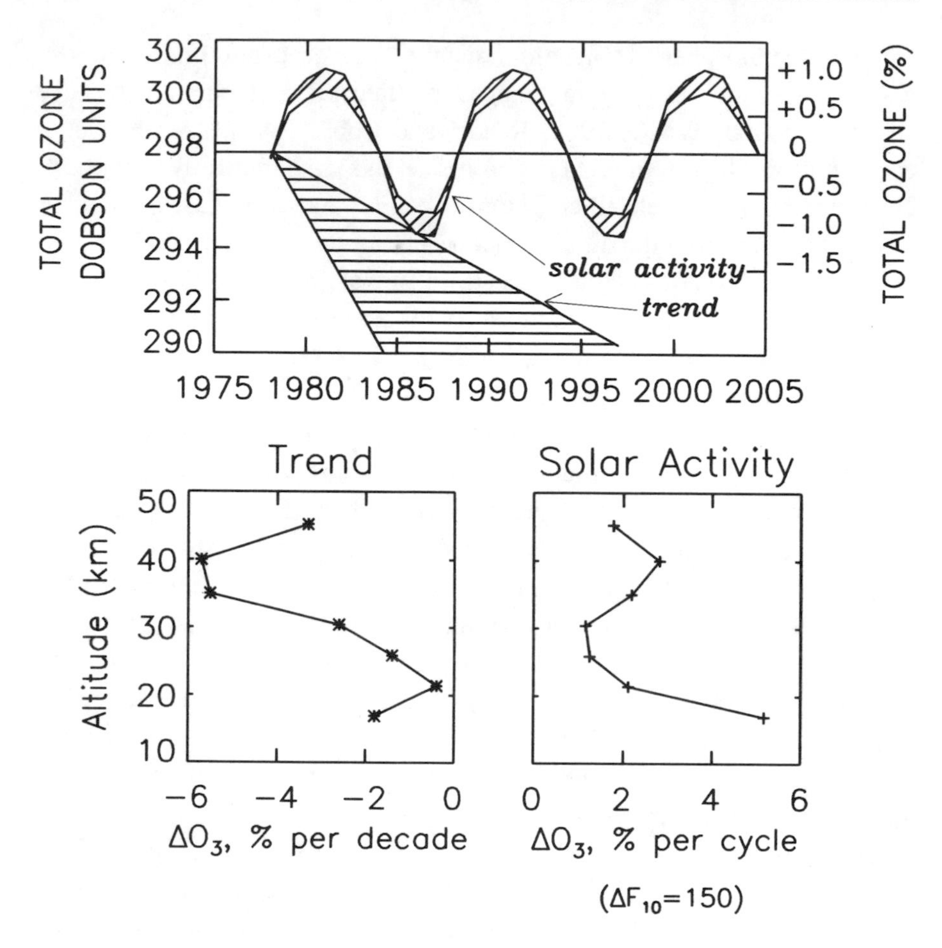

FIGURE 3.1 In the upper panel are compared changes in global column ozone concentrations in a vertical column above the Earth's surface during the 11-year solar cycle and the long term decrease derived from statistical analysis of the TOMS ozone measurements (Hood and McCormack, 1992). During decreasing solar activity from 1982 to 1986, solar forcing of ozone was approximately equivalent to the anthropogenic forcing, causing ozone depletion double that of the anthropogenic rate determined from the TOMS data, whereas from 1986 to 1991 the solar-induced changes approximately canceled the downward trend. Ozone changes were extended to solar cycles 22 and 23 by assuming equal levels of solar activity in all three cycles. Compared in the lower panels are altitude profiles of the mid-latitude Northern Hemisphere ozone decadal trend (left) and solar cycle variations (right) for a solar cycle with 150 units change of 10.7 cm flux (typical of cycles 21 and 22), deduced independently by Reinsel et al. (1994) from Umkehr data. Courtesy of J. Lean.

SOLAR ULTRAVIOLET RADIATION

Whereas total solar irradiance (discussed in Chapter 2) was, until recently, assumed to be invariant, solar emissions at shorter UV wavelengths are known to vary in phase with the solar activity cycle, with the shortest wavelengths varying the most (see Figure 1.1). The reason is that, compared with the visible radiation that contributes most to the total solar irradiance, the ultraviolet radiation is formed higher in the Sun's atmosphere and is more susceptible to the impact of magnetic fields that erupt into the solar atmosphere during times of enhanced solar activity. Accurate knowledge of the amplitude and temporal structure of solar UV spectral irradiance variations is clearly critical for studies of solar forcing of the middle atmosphere. The true extent of the UV irradiance changes over the 11-year solar activity cycle has long been debated and remains somewhat uncertain.

Estimates of the amplitude of the solar spectral irradiance variations shown in Figure 1.1 were derived from a combination of direct measurements and inferences from empirical variability models. Ultraviolet radiation is currently thought to vary over the 11-year solar cycle by 5 percent to 10 percent at 200 nm, 50 to 100 percent at H I Lyman α, and more than a factor of two at the shortest extreme ultraviolet (EUV) wavelengths, reaching maximum values during times of activity maximum, as indicated by the sunspot number. The solar cycle variations at wavelengths from 125 to 300 nm shown in Figure 1.1 are significantly less at all wavelengths than inferred directly from the rocket and satellite data base (Lean, 1987; Rottman, 1988), but still cannot be considered entirely reliable. Uncertainties remain even for the H I Lyman α line, whose variability is considered to be better known than for any other portion of the solar spectrum. Superimposed on the 11-year UV irradiance cycle is a 27-day rotational modulation whose amplitude near high activity levels may be as much as half the 11-year cycle amplitude.

Measurements of Solar UV Spectral Irradiance

Various instruments have been launched on rockets, satellites, and the Space Shuttle during the past 20 years (about two solar activity cycles) with the goal of acquiring knowledge of the Sun's ultraviolet spectral irradiance at wavelengths less than 400 nm. Solar Backscatter Ultraviolet (SBUV)-type instruments have been flown since November 1978, initially on the Nimbus 7 satellite (Heath and Schlesinger, 1986) and subsequently on a series of National Oceanic and Atmospheric Administration (NOAA) satellites (Donnelly, 1988; Cebula et al., 1992). Coincident with these observations from 1982 to 1989 are measurements by the solar spectrometer on board the Solar Mesosphere Explorer (SME) (Rottman, 1988; London and Rottman, 1990). Two solar spectral irradiance monitors for the region from 115 to 415 nm are currently operating on board the Upper Atmosphere Research Satellite (UARS), launched in September 1991 near the maximum of activity in solar cycle 22 (Rottman et al., 1993; Brueckner et al., 1993). Other solar UV spectroradiometers have been flown intermittently on rockets and balloons, with half a dozen recent measurements from the Space Shuttle.

Measurements from rockets (some of the more than 50) and spacecraft of the solar Lyman α irradiance are shown in Figure 3.2. The primary satellite data bases are the Orbiting Solar Observatory-5 (OSO-5), from 1970 to 1974 (Vidal-Madjar, 1975; Vidal-Madjar and Phissamay, 1980), Atmospheric Explorer-E (AE-E) from 1977 to 1980 (Hinteregger et al., 1981), SME from 1982 to 1989 (Rottman, 1988; Barth et al, 1990; White et al., 1990) and UARS, from late 1991 to the present (London et al., 1993); none overlap.

Figure 3.2 typifies, in many respects, the current state of the entire data base of solar irradiance at wavelengths shorter than of 400 nm. No portion of the spectrum has been monitored continuously for even one 11-year solar activity cycle. Instrumental effects contribute significantly to the scatter of the data. Absolute accuracies are typically quoted to be 10 to 50 percent, but inconsistencies between measurements made by different experimental groups are clearly apparent and signify larger systematic uncertainties. Information about instrument responsivity drifts during spacecraft missions is absent (in the case of AE-E), limited (in the case of SME), or speculative (in the case of SBUV). The SBUV diffuser reflectivity is thought to have degraded over seven years by about 30 percent at 250 nm (Herman et al., 1990). Although the SME instrument included a stored diffuser for periodic reference, this instrument suffered from problems associated with temperature-dependent wavelength calibration drifts. Pronounced long term drifts are also suspected in the AE-E data.

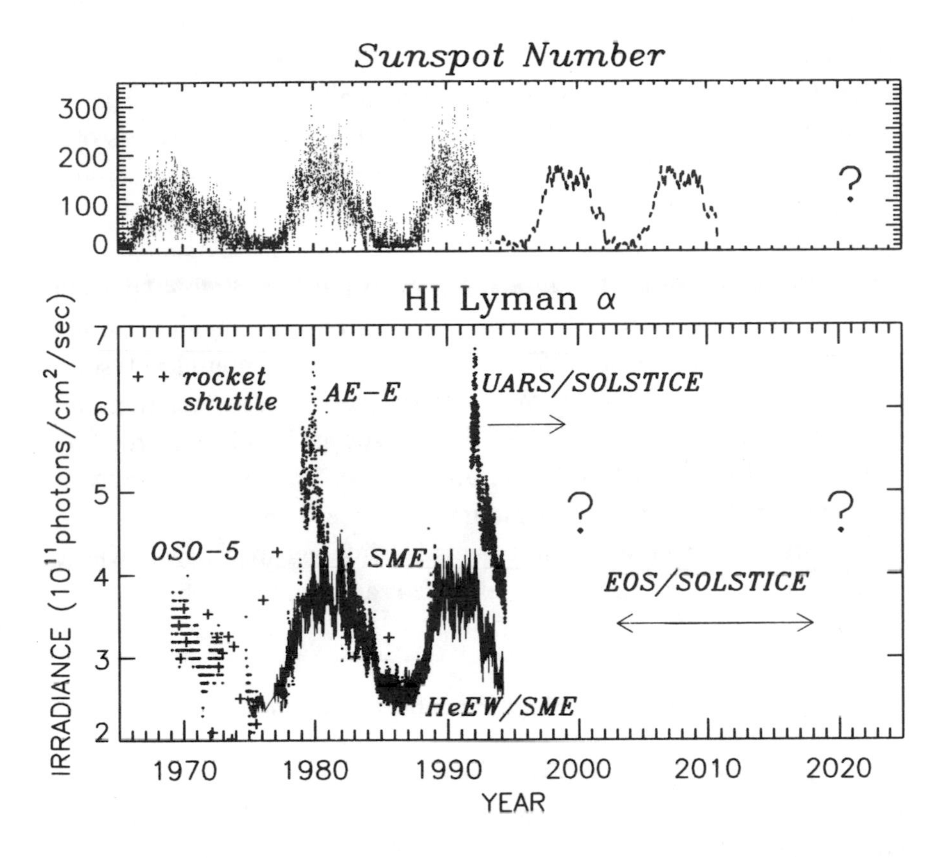

FIGURE 3.2 Contemporary solar activity variations as indicated by the sunspot number (top panel) and changes in the solar UV irradiance at Lyman α (bottom panel). The data (dots) are from OSO-5 (Vidal-Madjar, 1975), AE-E (Hinteregger et al., 1981), SME (from 1982 to 1989, Rottman, 1988), and UARS/SOLSTICE (London et al., 1993; courtesy of G. Rottman) and from rockets and the Space Shuttle (see Lean and Skumanich, 1983, for details). Also shown as a solid line is an extension of the SME data obtained from a linear relationship with the ground based He I EW (Figure 6.3). Note that for the Lyman α irradiance, $1\text{x}10^{11}$ photons/cm^2/sec is equivalent to about 1.63 mW/m^2. Courtesy of J. Lean.

Only the two UARS solar UV radiometers have the capability for end-to-end, in-flight sensitivity monitoring using, in the case of the Solar Ultraviolet Spectral Irradiance Monitor (SUSIM), a bank of four deuterium lamps and redundant optics, and for the Solar Stellar Irradiance Comparison Experiment (SOLSTICE), the UV fluxes from a collection of bright stars (Woods et al., 1993).

Despite the inadequacies in the experimental data base, there is no doubt that the Sun's full disk UV spectral irradiance varies. Changes of a few percent or more occur in response to the modulation of solar activity by the 27-day solar rotation and have been reliably measured at wavelengths as long as 250 nm. The amplitude of the rotational modulation is largest at the shortest wavelengths and can be as much as 30 to 40 percent at H I Lyman α (e.g., Lean, 1991). It has proven more difficult to extract from the data base reliable information about the magnitude of solar irradiance variations over the 11-year activity cycle. These variations are expected to be somewhat larger than those associated with solar rotation (Lean et al., 1992b), but are uncertain because of the degradation of instrument sensitivities during long-duration space flight, and because of the difficulty in comparing measurements by different instruments whose inaccuracies typically exceed the expected amplitude of the spectral irradiance variability.

Irradiance Variability Parameterizations

In lieu of an adequate experimental data base, information about the Sun's UV spectral irradiance variations has been acquired by critically analyzing the available data in conjunction with knowledge of solar activity derived independently, often from ground based observations (Lean et al., 1992b; DeLand and Cebula, 1993). In concert with rising solar activity during the 11-year cycle is an overall increase in the solar ultraviolet spectral irradiance because of increased emission from magnetic active regions whose radiation is enhanced at these wavelengths. Superimposed on this 11-year cycle are intermediate term variations over time scales of three to nine months that reflect the birth, growth, and decay of individual active regions associated with surges of activity. In addition, the UV irradiance is modulated by the Sun's 27-day rotation, which brings active regions onto the face of the Sun viewed from the Earth (e.g., Lean, 1987).

Irradiance variability parameterizations range from linear transformations of solar activity surrogates to detailed representations of active region sources of enhanced UV emission. Certain solar emissions that can be measured from the ground, such as the emission core of the Ca II K

Fraunhofer line, behave similarly to the UV irradiances in their responses to solar magnetic activity. These ground based data reflect the UV enhancement in active regions, exhibiting variations throughout the 11-year solar cycle in response to active region evolution and during the 27-day solar cycle. Estimates of solar cycle UV irradiance variability have been obtained either by extrapolating known 27-day changes, assuming a linear relationship with a suitable solar activity indicator (Heath and Schlesinger, 1986; Cebula et al., 1992; Lean et al., 1992b) or, in the case of Lyman α, by linearly regressing extant solar UV irradiance data against full-disk emission surrogates (e.g., White et al., 1990). Calculations of the UV spectral irradiance directly from ground based, spatially resolved Ca II K observations of the areas, locations, and spectral brightness of active regions have also been attempted with the purpose of elucidating the origins of the UV irradiance variations in terms of solar magnetic activity (Cook et al., 1980; Lean et al., 1982).

Inferences about the amplitude of the 11-year solar cycle variation in the UV spectral irradiance from 120 to 300 nm derived from measurements of the rotational modulation (Cebula et al., 1992; Lean et al., 1992b) are in general agreement with the variations reported from the SME observations (Rottman, 1988; London and Rottman, 1990). From 210 to 250 nm the solar cycle variation appears to be of the order of 3 to 5 percent, increasing to 7 to 9 percent at 200 nm. In the region of 120 to 200 nm the UV spectral emission lines vary significantly more than does the underlying continuum. For example, 1 nm spectral regions dominated by the H I Lyman α and O I 133.5 nm lines are estimated to vary by about 50 percent, with the underlying continuum varying by about half this amount.

However, parameterizations of solar UV irradiance variability derived from auxiliary solar data remain to be verified by more reliable and extended long term observations than SME, made by instruments such as those on UARS having complete in-flight sensitivity monitoring. This is true even for the Sun's strongest emission feature, the H I Lyman α line. Although its irradiance has been measured over more than three solar cycles, and far more frequently than observations of any other portion of the solar UV spectrum, unequivocal confirmation of Lyman α's true 11-year cycle variability has still proven difficult, confounded by apparent inconsistencies between different data sets and the suspicion of instrument artifacts in the data base. As demonstrated in Figure 3.2, none of the four Lyman α data bases overlap, which severely impedes the separation of the data into solar and instrumental effects. Data from the solar spectrometer on the SME satellite have been widely used for aeronomic applications, either directly for the period from 1982 to 1989, or over longer periods via parameterizations with the Ca II K (White et al., 1990) and He I 1083 nm EW (Lean, 1990) ground based indicators of chromospheric variability. However, measurements made by the AE-E (Fukui, 1990) and, more recently, the SOLSTICE on UARS imply significantly higher Lyman α irradiances (Figure 3.2) and greater variability than is estimated from parameterizations based on rotational modulation or SME observations. Pioneer Venus Orbiter Langmuir probe observations of the integrated solar EUV and Lyman α fluxes also provide indirect evidence for a Lyman α 11-year irradiance cycle in excess of 50 percent and poorly correlated with the Ca II and He I indices near the maximum of solar cycle 22 (Hoegy et al., 1993). Persistent inconsistencies among the measurements themselves and with parameterizations derived from solar activity surrogates caution against supposing that all is known even about the Sun's photon output variation at H I Lyman α or at any other UV wavelength.

ENERGETIC PARTICLES

Solar Proton Events

Eruptions of activity on the Sun, such as solar flares and coronal mass ejections, frequently give rise to fluxes of energetic particles. Interaction of these particles with the Earth's environment depends on the site of the eruption on the solar disk and on the particles' transport to the Earth via the solar wind and the Earth's near-space environment (discussed in Chapter 5). Typically, at 10 million electron volts (MeV), a large solar energetic-particle event reaches its peak at the Earth a few hours to a day after the initiating eruption and may persist for one to a few days. The particles are mainly protons and alpha particles with a small component of medium-mass and heavy ions. Fluxes at energies $>$ 10 MeV can reach 100,000 $cm^{-2}sec^{-1}sr^{-1}$.

Solar energetic particles are excluded from low magnetic latitudes by the Earth's magnetic field, but the polar regions are exposed to the full interplanetary flux. Figure 3.3 shows the regions in the two hemispheres (the polar caps) within which the effects are particularly felt, although during intense geomagnetic storms these boundaries can expand to considerably lower latitudes for brief periods. The depth in the atmosphere to which the particles penetrate depends on their initial energy; the energy spectrum of the particles is highly variable from event to event.

At times, solar protons can bombard the middle atmosphere at high latitudes for periods ranging from hours to days at a time. Penetration into the stratosphere (i.e., altitudes below about 50 km) requires energies in excess of 30 MeV for a proton, while direct penetration into the high-latitude troposphere requires energies greater than about 1 billion electron volts (GeV). Events with significant fluxes above this latter threshold are rare, but ground level neutron monitors, which detect thermal neutrons generated as secondary particles by the interaction of very energetic solar protons (several hundred MeV) with atmospheric constituents, have recorded many events.

The energy loss mechanism for solar protons in the Earth's atmosphere is almost exclusively through ionization. The impact between a solar particle and an atmospheric molecule releases a secondary electron with energy in the range of a few hundred electron volts that goes on to produce more ionization. Primary and secondary electrons both dissociate atmospheric nitrogen efficiently, while subsequent positive-ion reactions dissociate water vapor. The net result is the production of a wide range of odd-nitrogen (NO_x) and odd-hydrogen (HO_x) compounds, primarily in the middle atmosphere between heights of roughly 20 to 100 km. At heights of 60 to 70 km, where the solar energetic particle ionization reaches its maximum, it can exceed the normal background ionization rate by a factor of a million. These relatively large ionization rates give rise to significant increases in the concentration of free electrons and consequent severe disturbances in radio propagation at polar latitudes. In fact, the radio propagation effects provided much of the early information on solar energetic particle events before direct spacecraft measurements became available.

Perhaps one of the most surprising aspects of solar-terrestrial physics to emerge in the past decade has been the number and intensity of major

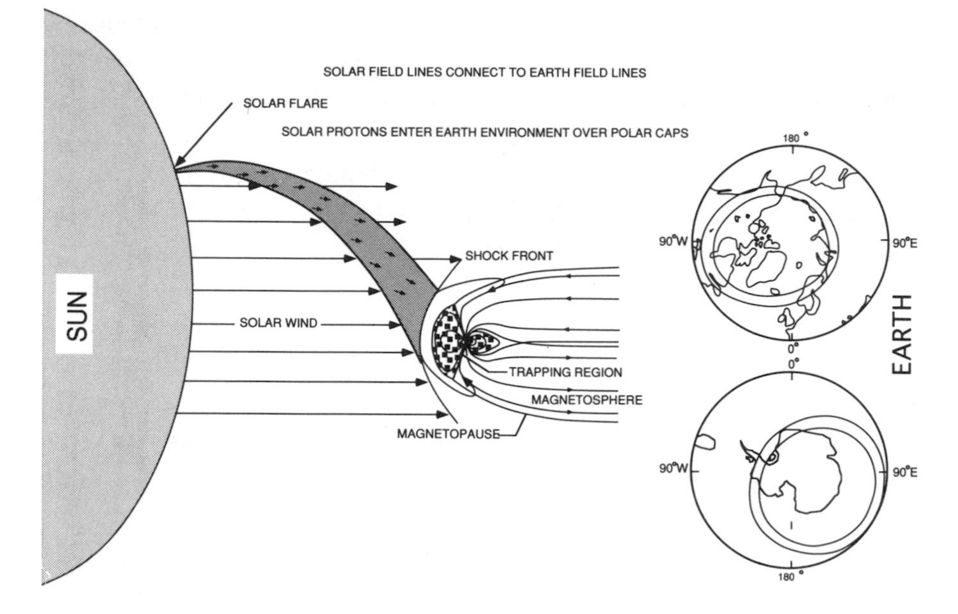

FIGURE 3.3 Shown schematically at left is the propagation of energetic solar particles from the Sun to the Earth. The interaction of solar and terrestrial magnetic field lines channels the particles toward the Earth at high latitudes, where they enter over the polar caps, as indicated in the polar diagrams on the right. The regions inside the inner curves experience virtually the entire interplanetary flux, while regions outside the outer curves are normally unaffected. Courtesy of C. Jackman and G. Reid.

solar particle events that can occur. Whereas the peak of solar cycle 20 (1969) showed substantial flaring and many moderate events, only a few of the particle disturbances (e.g., during August 1972) were considered really important from the point of view of solar-atmospheric coupling. In solar cycle 21, which peaked in 1979-1980, there were no flares as intense or as spectrally hard as the August 1972 event.

In solar cycle 22, however, the Sun provided many surprises. A number of exceptionally large eruptive events, comparable to the August 1972 event, occurred on the Sun during 1989. Very large events were seen in March and August, while in late September and early October an unusually intense and spectrally hard event occurred for which the flux of protons with energies > 10 MeV matched the highest values ever recorded. As well as wreaking havoc on sophisticated electronic systems

(Allen et al., 1989; Chapter 5), the solar protons associated with these major storms are thought to have led to signficant ozone depletions (Stephenson and Scourfield, 1991; Reid et al., 1991). Since September 1991, the Particle Environment Monitor (PEM) on board the UARS has been monitoring charged particle energy inputs to the middle and upper atmosphere (Sharber et al., 1993), including perhaps the most intense storm on record, that of November 1991.

Figure 3.4 provides a perspective for this recent activity by comparing the flux versus energy for the September 1989 event (cumulative proton dose at geostationary orbit for 29 September to 1 October above several energy thresholds) with the August 1972 total flux above 30 MeV. This 1972 point was essentially the entire significant flux for all of solar cycle 20 and was matched by the September 1989 point alone.

When all events are considered, solar cycle 22 emerged as one of the most active and most intensely disturbed periods known, with concomitant impact on the middle atmosphere.

Relativistic Electrons

Energetic electrons, like energetic protons, are capable of penetrating to the middle atmosphere. Enhanced relativistic electron fluxes (2 to 15 MeV) in the outer trapping zone of the magnetosphere have been observed at geostationary orbit since 1979. The presence or absence of multi-MeV electrons in the Earth's outer magnetosphere is controlled largely by high speed solar wind streams, which in turn show a strong solar cycle dependence related to the buildup of solar coronal hole structures in the declining phase of the 11-year sunspot cycle. The population dwindles as solar minimum is reached and is largely absent near the solar maximum. A 27-day periodic enhancement of the relativistic electrons is observed in association with concurrently measured solar wind streams in excess of 600 km/s.

The depth of penetration into the middle atmosphere of energetic electrons precipitated from the magnetosphere is a strong function of the electron energy (e.g., Baker, et al., 1987). However, the precipitating electron energy spectrum and its flux at the top of the atmosphere for relativistic electron precipitation (REP) events are still uncertain. A 1 MeV electron can penetrate to about 55 km, while a 10 KeV electron only

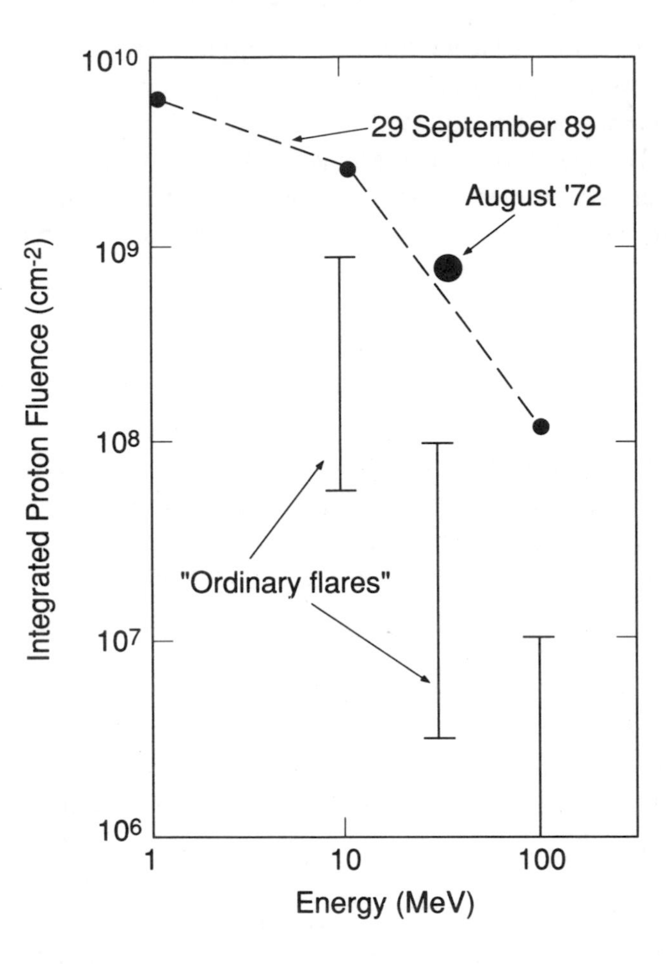

FIGURE 3.4 The integrated proton flux of high energy solar protons typical of ordinary flares is compared with two of the largest disturbances seen in the spacecraft era, in August 1972 and September 1989. Heath et al. (1977) reported significant changes in ozone composition as a result of the August 1972 event. Reid et al. (1991) have analyzed the effects on the middle atmosphere of solar proton events during August-December 1989. Courtesy of D. Baker.

penetrates to about 100 km. Thus, if there is a significant population and flux of trapped high-energy relativistic electrons (energy $>$ 700 KeV) in the magnetosphere, precipitation in auroral and subauroral latitudes may occur, with energy deposition and resulting production of hydrogen oxides and nitrogen oxides in the lower mesosphere and upper stratosphere. If the energetic events are sufficiently frequent, the downward transport of nitrogen oxide molecules during the winter and spring could result in a significant effect on the budget of odd-nitrogen in the lower stratosphere (e.g., Dahe et al., 1991).

Measurements from the Spectrometer for Energetic Electrons (SEE) on spacecraft 1979-053 and 1982-019 do provide evidence for large fluxes of relativistic electrons at 6.6 Earth radii (R_E) that undergo variations on 27-day and 11-year time scales (Baker, et al., 1987). Precipitation of 30 to 40 percent of these electrons from the largely trapped trajectories seen at synchronous orbit into the atmosphere could have significant effects on middle atmosphere chemistry (Baker et al., 1993). However, measurements of the energy deposited into the middle atmosphere made by sounding rockets (e.g., Goldberg et al., 1984) have typically indicated much smaller fluxes of relativistic electrons than measured by the SEE instrument. Figure 3.5 shows a comparison of a typical energy deposition rate from an REP event measured by Goldberg et al. (1984) with energy deposition rates of EUV and galactic cosmic rays. In the lower mesosphere and stratosphere the energy deposition rates are almost two orders of magnitude larger for the Baker et al. REP event than for the Goldberg et al. event. On the other hand, recent rocket observations (Herrero et al., 1991) indicate that the fraction of the high altitude flux of relativistic electrons that reaches the middle atmosphere near noon (when daytime REP fluxes peak) may actually be significantly larger, and penetrate deeper, than the nighttime REP events measured by earlier rockets (Goldberg et al., 1984). At about 100 km, electron spectra were estimated to be 5 to 25 percent of the geostationary orbit fluxes at the same local time, and Baker et al. (1993) argue that precipitated fluxes of electrons with energies > 1 MeV could be as much as 30 to 50 percent of the trapped flux levels. New measurements being made by the Solar, Anomalous, and Magnetospheric Particles Explorer (SAMPEX) may help to resolve these differences and determine whether REPs are as important for long term ozone variability as has been inferred from some studies (Callis et al., 1991).

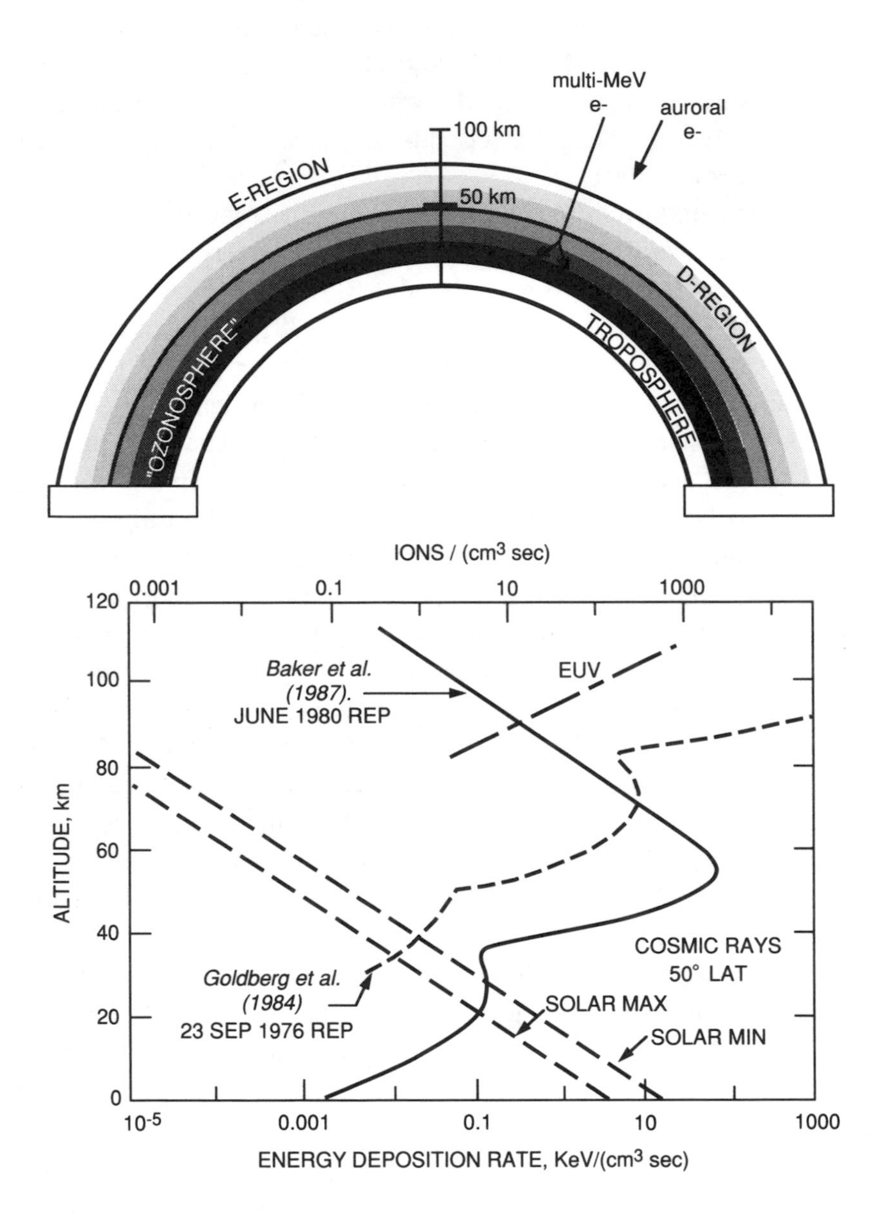

FIGURE 3.5 The upper figure depicts the penetration of energetic particles in the middle atmosphere. In the lower figure, energy deposition due to relativistic electrons in June 1980 and September 1976 are compared with energy deposition from extreme ultraviolet radiation and galactic cosmic rays during minima and maxima of the 11-year activity cycle. Courtsey of Jackman (1991), adapted from Baker et al. (1987) and Goldberg et al. (1984). Copyright by the American Geophysical Union.

Galactic Cosmic Rays

The Sun is not the only source of energetic particles that penetrate the Earth's atmosphere. Galactic cosmic rays (GCR), which are mostly protons and alpha particles, have typically higher characteristic energies (peaking in the range 0.1-1 GeV per nucleon near Earth orbit) but lower fluxes than the solar (cosmic ray) protons. Carbon, nitrogen, oxygen, and other medium-mass nuclei are also present, together with a significant heavy ion population. Like solar energetic protons and relativistic electrons, galactic cosmic rays are strongly shielded from reaching the equatorial regions of the Earth's surface by the terrestrial magnetic field and by the thick neutral atmosphere. This shielding is energy dependent, deflecting incident particles with energies as high as 15 GeV at the magnetic equator. However, in the polar regions of the Earth and at high altitudes (> 5 Earth radii) the GCR component has direct access to the middle atmosphere.

At energies less than about 1-2 Gev, the main mechanism for energy loss in the atmosphere is through ionization, and the particles stop well above the surface. At higher energies, nuclear interactions with atmospheric constituents become increasingly more important, giving rise to energetic secondary particles that can penetrate to the surface of the Earth. Among the products of these interactions are neutrons, which are created with relatively high energies and can subsequently be thermalized as they travel down through the atmosphere. The fact that these neutrons can be readily measured at the Earth's surface provides a convenient means of monitoring cosmic ray fluxes on a routine basis.

There is a relatively strong (5 to 15 percent) solar cycle modulation near Earth's orbit for galactic cosmic rays. This modulation, which is also energy dependent and increases at lower energies (below 1 GeV) to well over a factor of two, is in antiphase with the 11-year solar activity (sunspot) cycle and reflects the effects of changes in magnetic structures in the solar wind on the flow of cosmic rays into the solar system. Significant cosmic ray (Forbush) decreases also occur during large magnetic storms associated with coronal mass ejections and usually with solar flares (see Herman and Goldberg, 1978; Tinsley and Deen, 1991).

Cosmic ray secondary neutrons have an important by-product in the form of ^{14}C, or radiocarbon, produced by the reaction of a neutron with

^{14}N, the most abundant atmospheric isotope. The ^{14}C becomes incorporated into the terrestrial carbon cycle and eventually into the carbon of living organisms. Once created, the ^{14}C decays with a half-life of 5730 years, thereby providing the basis for the radiocarbon dating technique that has revealed much about biosphere evolution on millennial time scales. Laboratory measurements of the ^{14}C:^{12}C ratio in tree-ring dated wood have provided a chronicle of ^{14}C input variations and hence of variations in solar activity that now extend almost 10,000 years into the past (e.g., Stuiver and Braziunas, 1993).

SOLAR FORCING OF THE MIDDLE ATMOSPHERE

Changes in both solar UV radiation and in the flux of energetic particles influence the middle atmosphere directly. Determining the impact of human activities on upper stratospheric ozone and temperature requires a knowledge of the impact of this solar forcing background.

Effects from Variations in UV Irradiance

As discussed previously, variations in the solar flux during the 11-year activity cycle are still not well understood, particularly at the ultraviolet wavelengths that are responsible for the primary production and destruction mechanisms of middle atmosphere ozone. Changes in ozone and temperature resulting from the 27-day solar irradiance rotation cycle are better understood (Hood, 1987; Hood and Douglass, 1988; Chandra, 1991; Brasseur, 1993) and have confirmed the response of the middle atmosphere to changes in solar UV energy inputs. Ozone changes related to the 11-year irradiance cycle are expected to be larger than those detected over solar rotation time scales (Chandra, 1991) because the UV irradiance variation is larger (Lean, 1991) and the sensitivity of ozone to longer period forcing is greater (Brasseur, 1993).

Trend studies of middle atmosphere ozone and temperature observations provide evidence for long term changes associated with solar cycle variations in UV radiation, albeit with large uncertainties in the magnitude of the effect, which has thus far been measured for only one 11-year solar cycle. The results indicate that over decadal time scales solar cycle forcing is a sizable contributor to trends in global ozone and upper stratospheric temperatures. This is illustrated in Figure 3.1, where the solar cycle in global total ozone is compared with the long term trend deduced from the TOMS observational record. Time series statistical analyses of the TOMS data (corrected for instrument degradation) indicate solar cycle changes in globally averaged total ozone of the order of 1.8 $\pm$ 0.3 percent, in phase with the activity cycle (Stolarski et al., 1991, Chandra, 1991, Hood and McCormack, 1992). Ground based data signify a slightly smaller effect, in the range 1.2 $\pm$ 0.4 percent to 1.5 $\pm$ 0.9 percent (Reinsel et al., 1988; Bojkov et al., 1990). The trends are larger at higher latitudes than in the tropics (Randel and Cobb, 1994; Reinsel et al., 1994a).

Both spacecraft and ground based observations indicate that significant solar cycle changes are also occurring in the vertical ozone profile. Global upper stratosphere (about 45 km) ozone is estimated from SBUV data to increase by 3 to 4 percent from the minimum to the maximum of the 11-year solar cycle (Hood et al., 1993). Similar results were found by Reinsel et al. (1994b) using northern hemisphere mid-latitude Umkehr data (revised for aerosol effects). Changes of this magnitude may also be occurring in the lower stratosphere (about 20 km), with a reduced solar cycle effect of about 1 percent near 30 km. For comparison, the data indicate upper stratosphere depletion at northern mid-latitudes of the order of 6 percent per decade and about 3 percent per decade in the lower stratosphere (Reinsel et al., 1994b), which thus may be masked by solar-cycle effects for a number of future solar cycles (Figure 3.1). In addition, solar cycle variations have been reported in both stratospheric temperatures and winds (Mohanakumar, 1989; Kodera and Yamazaki, 1990; Chanin and Keckhut, 1991; Hood et al., 1993).

That changing UV radiation throughout the 11-year solar activity cycle should evince a measurable effect on upper stratospheric ozone and temperature has been predicted by atmospheric modeling studies over the past decade. Assuming an irradiance increase at 205 nm from solar minimum to solar maximum ranging from 6 to 9 percent based on available spacecraft data, and using H I Lyman α and He I EW surrogates for the variations in UV flux with time, Wuebbles et al. (1991) determined from a two-dimensional chemical-radiative-transport model of the global middle atmosphere a corresponding increase in total ozone of 1.2 to 1.7 percent. The largest local changes, about 2 to 3 percent, occurred at about

40 km altitude, while maximum temperature changes, 0.75 to 1 K, were determined near the stratopause. Contrary to the observations, significant solar-induced ozone changes in the lower stratosphere are not simulated by two-dimensional middle atmosphere models; equally small changes were found in a 2D model study by Huang and Brasseur (1993). Furthermore, when downward transport of the upper atmosphere nitric oxide (NO) enhancement at solar cycle maximum is included in 2D model simulations, there is significant ozone depletion, resulting in an antiphase global total ozone change, also contrary to the observations.

Clearly, substantial uncertainties exist in determining the response of middle atmosphere ozone and temperature to solar radiative forcing. In part, this is because the magnitude of the solar flux variations at the important wavelengths (200-300 nm) are still uncertain. Perhaps more importantly, a primary limitation of the 2D models is that dynamical feedbacks are not available to amplify initial perturbations of the middle atmosphere by changing UV energy inputs (Balachandran and Rind, 1994). To clarify the discrepancies between observed and modeled ozone response to solar forcing it is essential to continue to monitor the relationships between variations in the middle atmosphere parameters and observed solar irradiance variations, both for the 27-day rotation and the 11-year solar cycle, as well as longer term trends. Because the prime space based global data sets extend over only one 11-year solar activity cycle, more definite studies of, for example, the role that might be played by dynamics and by the QBO in the observed solar cycle variation of ozone, await longer observational records.

Effects from Solar Proton Events

During a solar proton event, the primary energetic protons produce secondary electrons by impact on atmospheric molecules, and these electrons can efficiently dissociate molecular nitrogen, producing a range of odd-nitrogen species by subsequent chemical reactions. The odd-nitrogen, in turn, catalytically destroys ozone in the stratosphere. This affects the Earth's surface environment directly through an increase in the solar ultraviolet flux and may have indirect effects via coupling of the middle atmosphere to the troposphere. Ozone is also depleted at higher altitudes in the mesosphere as a consequence of the dissociation of water vapor, producing odd-hydrogen that destroys ozone at these altitudes through a catalytic reaction sequence.

The odd-nitrogen and odd-hydrogen effects on ozone have been observed directly by satellites and rockets during solar proton events (e.g., Weeks et al., 1972; Crutzen et al., 1975; Heath et al., 1977; Thomas et al., 1983) and have been studied theoretically by several groups (e.g., Jackman et al., 1990; Reid et al., 1991). As demonstrated in Figure 3.4, the August 1972 solar proton event was one of the largest events in the past 30 years. Substantial increases in nitrogen oxides were expected to be associated with this event. Available satellite measurements and modeling studies indicate ozone depletions >20 percent in the upper stratosphere, with depletions >15 percent persisting for about two months after the event, indicating a slower recovery time than is estimated by two-dimensional modeling studies (Jackman et al., 1990).

In a sense, each major solar proton event can be looked on as a controlled experiment, in which a calculable amount of NO_x is introduced into the atmosphere, and the subsequent effects can be monitored. These events are thus a potentially valuable source of information on the behavior of the atmosphere, providing a unique means of testing the validity of model predictions (Jackman and McPeters, 1987; Jackman, 1991). Also, over the longer time scales of the 11-year solar cycle, variations in the downward transport of nitrogen oxides produced by auroral particle precipitation at altitudes in the thermosphere (Garcia et al., 1984) and other solar particle effects could further affect stratospheric temperature and ozone. Understanding these processes, however, requires accurate monitoring of the particle influx at the top of the atmosphere, since this is the basic information needed to calculate the NO_x production. Since the most relevant effects are those occurring at the lowest levels in the atmosphere, the high-energy spectrum must be reliably known beyond energies of 100 MeV. Routine spacecraft monitoring of energetic-particle fluxes has generally been carried out only for energies much less than this.

Effects from Relativistic Electron Precipitation

Various studies have suggested that relativistic electron precipitation events may be important contributors to the odd-nitrogen budget of the middle atmosphere (Thorne, 1980; Baker et al., 1987; Callis et al., 1991).

Thorne (1980) indicated that the odd-nitrogen and odd-hydrogen formed in the upper stratosphere and lower mesosphere by the ion chemistry associated with these precipitation events can lead to destruction of ozone. Although these studies suggest that most of the production of the odd-nitrogen from such events occurs at subauroral latitudes in the lower mesosphere, the downward polar transport during the winter and spring may provide a source of stratospheric nitrogen oxides. Callis et al. (1991) have suggested that this phenomenon may explain a significant fraction of the measured lower stratospheric ozone decrease in the 1978-1986 period. However, as discussed earlier, there remain many uncertainties associated with the amount of nitrogen oxides produced and the resulting effects on lower stratospheric ozone.

ULTRAVIOLET RADIATION REACHING THE BIOSPHERE

The amount of solar ultraviolet radiation reaching the biosphere is extremely sensitive to the amount of overhead ozone, especially at the UV-B wavelengths that coincide with a sharp decrease in the ozone (O_3) absorption at about 310 nm (Figure 1.2), the edge of the O_3 Hartley band. Therefore, the amount of ozone in the middle atmosphere, and its response to natural and anthropogenic influences, is extremely important for determining potential impacts on the UV-sensitive biosphere (Madronich, 1992). Small reductions in ozone column density are expected to cause an increase in the incidence of skin cancer (Madronich and de Gruijl, 1993) and in blindness due to UV-B induced cataracts, as well as decreased agricultural and fisheries production. Biological immune systems, as well as the oxidizing capacity of the troposphere, are also expected to be affected.

Although measurements indicate that total ozone has been decreasing over the past two decades, and theoretical calculations predict a corresponding increase in the UV radiation reaching the Earth's surface (Madronich, 1992), experimental evidence for this increased UV exposure is at present inconclusive. Factors such as tropospheric ozone, sulfate particles in the atmosphere, and cloudiness can produce opposite effects (Crutzen, 1992). Substantial downward, upward, and null trends have all been reported (Justus and Murphey, 1994). At the same time, there are questions about the adequacy of the primary device used in these measurements, the Robertson-Berger meter. The measurements require not only accurate absolute radiometry, but also high wavelength resolution and accuracy, because the ozone absorption spectrum at the edge of the Hartley band is so steep. The spectral intensity of solar radiation received at Earth drops by eight orders of magnitude in the spectral band from 300 to 280 nm. The data from the existing ground based network probably lack adequate calibration stability, limiting their suitability for trend analyses.

4

Solar Variations and the Upper Atmosphere

BACKGROUND

Extending from the top of the middle atmosphere to some hundreds of kilometers into space is the Earth's upper atmosphere (Figure 1.2) and its embedded ionosphere. This tenuous layer of neutral and charged particles shields the human habitat from high energy solar radiation and particles, enables part of the extensive communication network on which society increasingly relies, and is the medium in which thousands of spacecraft now orbit. Unlike the relatively placid lower atmosphere, the upper atmosphere is a region of extreme spatial and temporal variability, constantly agitated by solar radiative and auroral forcings. Driving the processes that at any instant define the physical state of the upper atmosphere and ionosphere is the solar radiation at wavelengths less than about 180 nm. Many of the region's continually changing physical phenomena derive directly or indirectly from changes in this radiation and from the impact of energetic particles channeled into the upper atmosphere at high latitudes via the Earth's magnetic field.

While solar variability exerts a dominating influence on the Earth's upper atmosphere, any direct effect on the biosphere appears to be more subtle than that exerted by solar forcing of the middle and lower atmospheres. The fact that the highly variable upper atmosphere is coupled to the middle atmosphere through chemical, radiative, and dynamical mechanisms, and to the troposphere through the global electric circuit, cannot be ignored. Understanding how the upper atmosphere varies naturally, and how it may be affected by human activities, is necessary from a societal and economic perspective because of the critical role played by the upper atmosphere in communications, navigation, national defense, and a wide assortment of space related endeavors, including the presence of humans in space. Furthermore, current modeling studies indicate that the upper atmosphere may itself be sensitive to global change caused by human activities.

SOLAR EUV AND UV RADIATION

The Sun's ultraviolet radiation at wavelengths less than about 180 nm varies considerably more than does the UV radiation that is absorbed in the middle atmosphere and the visible radiation that penetrates to the Earth's surface (see Figure 1.1). Solar cycle changes of 100 percent are typical in solar radiation at wavelengths from 10 to 100 nm; the soft X-rays (1 to 10 nm) that penetrate to the lowest layers of the upper atmosphere vary by an order of magnitude. This highly variable energy from the Sun is deposited entirely in the terrestrial upper atmosphere via absorption of the primary constituents, O_2, N_2, and O. Without heating from the absorption of solar extreme ultraviolet (EUV) and UV radiation, the thermosphere and the ionosphere would not exist at all. This heating, which varies with solar activity, is responsible for the increase of temperature with height above about 100 km (see Figure 1.2) and for driving most of the bulk motions of the gases within the entire region. Large variability in the basic properties of both the thermosphere and ionosphere is the direct result of the variability in the solar EUV and UV input (as illustrated in Figure 1.2 by the change in the temperature profile from minimum to maximum solar activity).

Measurements of Solar EUV Spectral Irradiance

Current knowledge of the magnitude and variability of the solar EUV energy deposited in the upper atmosphere is based almost entirely on a brief four-year period of measurements made by the Atmosphere Explorer

(AE-E) spacecraft about 15 years ago. These measurements revealed a considerable increase in the solar EUV flux during the ascending phase of solar cycle 21. Some emissions at wavelengths shorter than 30 nm increased by factors of 10 to 100 between solar minimum conditions in 1976 and maximum activity in 1980. These emissions emanate from the highest, hottest layers of the Sun's atmosphere (the solar corona). Radiation at wavelengths between 30 and 120 nm, formed lower in the solar atmosphere (the chromosphere), varied somewhat less, by factors of two to three from the minimum to the maximum of activity in solar cycle 21. At still longer UV wavelengths, solar cycle variability decreases from a factor of two near 100 nm to about 10 percent near 200 nm. In addition to the overall change in solar radiation between solar minimum and maximum, the AE-E data showed shorter term fluctuations on a monthly, daily, and even an hourly basis, with the coronal emissions being much more variable than the chromospheric emissions.

Essentially all interpretive studies of upper atmosphere phenomena now use scenarios of solar variability derived from the AE-E data base. However, AE-E did not monitor the highly variable soft X-rays, nor do the AE-E data agree with earlier rocket measurements about either the magnitude or the variability of the EUV irradiance (Lean, 1988). Concerns about the validity and limitations of the AE-E data base continue to be raised. AE-E's absolute irradiance calibration was derived from two Air Force Geophysics Laboratory rocket measurements, one during 1974 (which preceded the AE-E data) and another in 1979 (Heroux and Hinteregger, 1978; L. Heroux, private communication, 1981). Possible changes in the sensitivity of the AE-E instruments throughout the mission are unknown, since no provision was made for in-flight calibration. A comparison of the 1979 rocket measurement used for the AE-E calibration with a recent rocket measurement (Woods and Rottman, 1990) indicates significant inconsistencies in that only the strongest emission lines were enhanced in the 1979 spectrum, for which solar activity levels were higher. This contradicts current understanding of the origin of the EUV irradiance variations, which predicts that solar activity causes an increase in the EUV radiation at all wavelengths. The discrepancy is most likely the result of instrumental effects (see Lean, 1990 for details).

AE-E ceased operation at the end of 1980. In the ensuing decade only a few isolated measurements of the solar EUV spectral irradiance were made -- from the San Marco satellite for nine months in 1988 (Schmidtke et al., 1992) and from a few rockets (Woods and Rottman, 1990). Some additional measurements of the EUV radiation integrated over very wide spectral bands have also been made from rockets (Feng et al., 1989; Ogawa et al., 1990). None of these observations has succeeded in clarifying the true amplitude and variability of the solar EUV radiation. Like the AE-E observations, they are compromised by an assortment of instrumental effects that make it extremely difficult to extract true solar spectral variability.

Since its launch in mid-1991, the Yohkoh satellite (Petersen et al., 1993) has monitored the soft X-ray flux from the Sun for most of the descending portion of solar cycle 22, providing uniquely valuable data about a highly uncertain region of the solar spectrum. Continuing these observations into the upcoming solar minimum and subsequent activity maximum will contribute to improved understanding of solar forcing of lower thermospheric NO concentrations, and the possible transfer of chemical energy between the upper and middle atmospheres.

Irradiance Variability Parameterizations

The absence of continuous, reliable observations of the solar EUV spectral irradiance has forced reliance on empirical variability models based on solar activity surrogates to estimate EUV spectral irradiances for use in upper atmosphere research and in operational applications. The AE-E solar irradiance data have been used to construct parameterizations of the solar EUV flux variations as a function of primarily the solar 10.7 cm radio emission that can be measured from the ground (Hinteregger et al., 1981; Tobiska, 1991). The measured solar EUV flux values cover the period from 1976 to 1980, but the solar flux models have been used to represent the solar EUV and UV fluxes for other periods, generating values that are typically inconsistent with earlier data (Figure 4.1). Furthermore, different empirical models developed from ostensibly the same data base can predict quite different EUV spectral irradiances (Lean, 1990).

Aeronomic studies of thermospheric and ionospheric properties indicate, not surprisingly, that existing solar EUV irradiance variability models are inadequate for many geophysical applications. While considerable

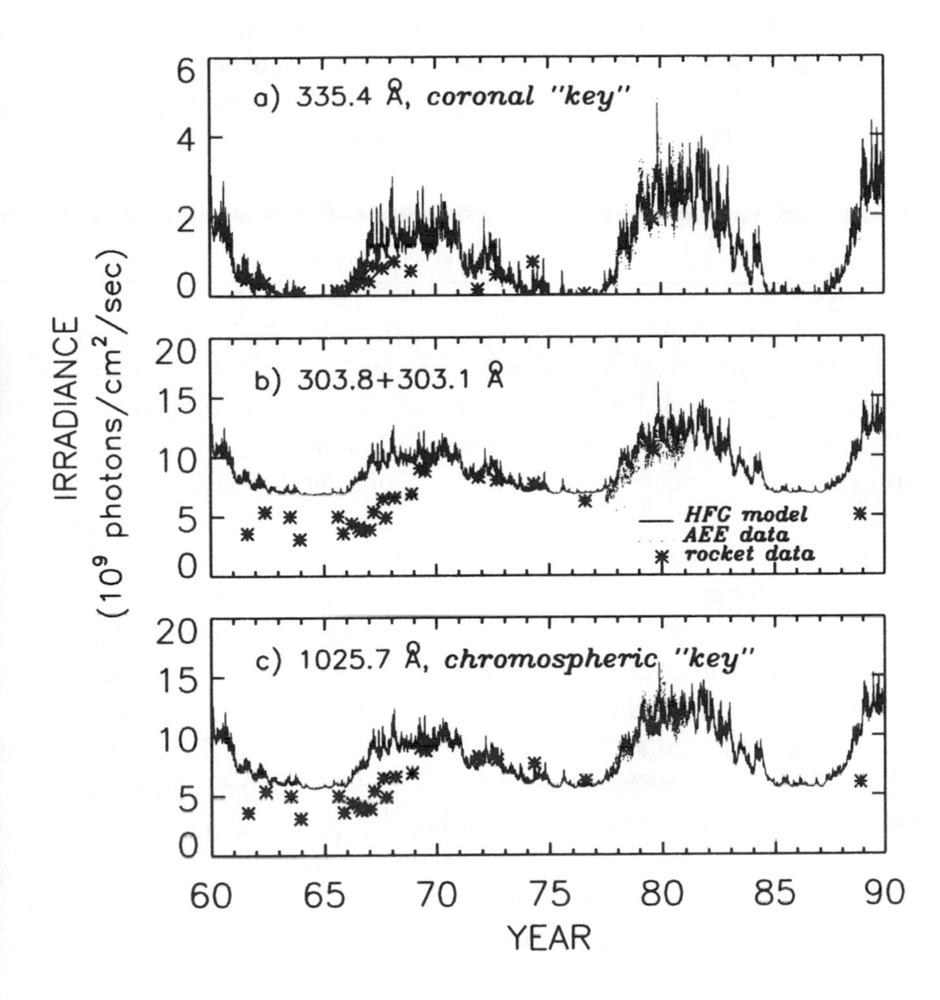

FIGURE 4.1 Comparison of measured solar EUV irradiance variations with the empirical model calculations of Hinteregger et al. (1981) based on the 10.7 cm flux. The variations are shown for a) the coronal emission at 33.54 nm, b) the primarily chromospheric emission at 30.38 nm, which is the singularly most important solar emission line for heating the Earth's upper atmosphere, and c) the chromospheric line at 102.57 nm. The model calculations (solid line) are based on the AE-E data (dots) in solar cycle 21 and do not show very good agreement with rocket measurements (asterisks) during the previous solar activity cycle. Adapted from J. Lean, Advances in Space Research, 8, (5)263, 1988, with permission from Elsevier Science Ltd, Pergamon Imprint, The Boulevard, Langford Lane, Kidlington 0X5 1GB, UK, and J. Lean, J. Geophs. Res., 95, 11939, 1990, copyright by the American Geophysical Union.

progress has been made during the past decade in developing sophisticated models of thermospheric and ionospheric aeronomic processes and global dynamics, there has been little improvement in the (currently inadequate) empirical parameterizations of solar EUV and UV radiation that these models use. This has led to the application of various correction factors to the original data to achieve agreement between model predictions and aeronomic observations. For example, the solar EUV flux below 20 nm has been doubled to force agreement with observed photoelectron spectra (Richards and Torr, 1988; Winningham et al., 1989) and the flux below 5 nm has also been scaled upwards to account for a measurement of thermospheric NO at maximum levels of solar activity (Siskind et al., 1990).

Improved solar irradiance variability models are needed not just for aeronomic research but, increasingly, for operational applications such as forecasting ionospheric conditions (Balan et al., 1994) and for predicting thermospheric density for satellite orbital and point density determinations. The lifetime and utility of an Earth-orbiting object depend on the density structure of the upper atmosphere, which is controlled by solar EUV radiation and consequently varies over time scales from hours to decades (White et al., 1994). Desired accuracies of 5 percent for thermospheric densities for operational purposes require a similar accuracy in knowledge of the solar EUV and UV spectral irradiance. The need for this knowledge is demonstrated in Figure 4.2, where the orbital decay rate of the SMM satellite can be seen to track changes in solar activity (as indicated by the 10.7 cm radio flux). SMM's reentry into the Earth's atmosphere is thought to have been accelerated by the progressively increasing solar activity in 1989, the ascending phase of solar cycle 22. High uncertainty surrounded the launch of the Hubble Space Telescope because of insufficient knowledge of the atmospheric drag that it would experience when launched at a time near maximum solar activity (Withbroe, 1989).

AURORAL PARTICLE AND ELECTRIC FIELD INPUTS

The Earth's thermosphere and ionosphere system responds not only to changes in the solar EUV and UV radiative input but also to particulate inputs of solar energy and momentum at high latitudes associated with auroral processes. Although these inputs are dominant at high latitudes

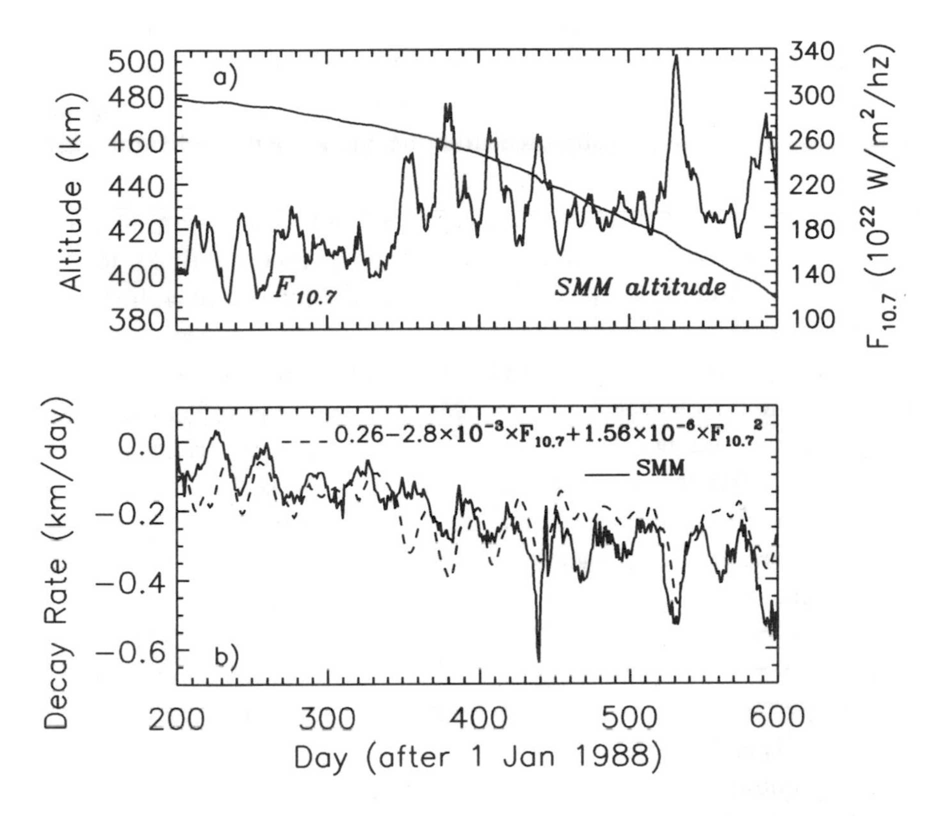

FIGURE 4.2 Shown in the upper panel is the decreasing altitude of the Solar Maximum Mission (SMM) satellite (D. Messina and G. Share, private communication, 1990) as a function of time just prior to its reentry into the Earth's atmosphere in December 1989, coincident with increasing solar activity as indicated by the Sun's 10.7 cm radio flux, $F_{10.7}$. The lower panel shows that the orbital decay rate (solid line), determined as the change per day in the altitude, is strongly influenced by variations in solar energy input, as indicated by the daily $F_{10.7}$ solar activity proxy (linearly transformed to an equivalent decay rate, dashed line). The cycle of about 27 days occurs because the Sun's rotation causes active regions to move across the face of the solar disk seen at the Earth, modulating its output of UV and EUV radiation. When the Sun's radiation is brightest, the Earth's atmosphere expands outwards and the rate of decay of the satellite orbit increases. Active regions that cause enhancements of the UV radiative output also modify the 10.7 cm radio flux. From J. Lean, Reviews of Geophysics, 29, 511, 1991, copyright by the American Geophysical Union.

(Figure 3.3), they have a great variability that affects the basic structure and dynamics of the entire upper atmosphere system. At times, such as in brief periods during intense geomagnetic storms, energized electrons bombard the upper atmosphere, colliding with atmospheric constituents and transferring their energy, resulting in visual displays of auroral phenomena. The energy deposited at high latitudes in the aurora can increase by as much as two orders of magnitude relative to geomagnetically quiet conditions, locally exceeding the energy deposited from solar EUV radiation. Auroral energy inputs are known to have a significant effect on aeronomic processes and dynamics of the ionosphere, thermosphere, and mesosphere and perhaps indirectly (via couplings to the stratosphere) on the troposphere, even though the physical couplings are not understood.

In addition to knowledge of radiative energy inputs, global dynamic models of the thermosphere and ionosphere system require knowledge of the global distributions of auroral particle precipitation, electric fields, and currents. During the past decade, spacecraft such as the Atmospheric Explorer and Dynamics Explorer, as well as various ground based programs, have provided a good first order understanding of the energy inputs to the thermosphere and ionosphere. Some information on global particle inputs has been derived from satellite images of UV and visible auroral airglow. However, many unresolved questions remain about the variability of the fundamental energy inputs and the global distribution of electric fields and currents.

Many questions also remain about the impact of auroral processes on global change. For example, how are the atmospheric chemical species such as NO that are produced by auroral processes transported globally? How might they be transported to the lower atmosphere, where they may influence global atmospheric properties? How do the enhanced currents and fields produced during geomagnetic storms influence properties of the troposphere and at the ground by coupling into the Earth's global electric circuit?

GLOBAL CURRENTS AND ELECTRIC FIELD COUPLINGS

The Earth and its atmosphere are almost permanently electrified. The Earth's surface has a net negative charge, and an equal positive charge is distributed throughout the atmosphere. The atmospheric region above about 60 km is generally considered the upper conducting boundary of the global electric circuit, which includes electrical interactions between all atmospheric regions. This upper conducting boundary is formed by solar ionization of atmospheric constituents.

Global Circuit Processes

Three main generators operate in the Earth's global electric circuit (Figure 4.3): (1) thunderstorms, (2) the ionospheric wind dynamo, and (3) the solar wind/magnetosphere dynamo. These processes are reviewed in NAS (1986) and are only briefly described here.

Thunderstorms are electrical generators whose global activity provides a current output that maintains a vertical potential difference of about 300 kV between the ground and ionosphere, with a total current flow between the two of about 10^3 A. The variability of thunderstorm activity results in diurnal, seasonal, and interannual variations in the potential differences and currents in the circuit. The electrical processes associated with thunderclouds are many and complex. In general, a conduction current flows from the top of the cloud toward the ionosphere and into the global circuit. Beneath the thundercloud a number of complex currents flow. The total Maxwell current, defined as the sum of all of these current systems plus the displacement current, has been shown to vary slowly over the storm's history, suggesting that this electrical quantity is coupled to the meteorological structure of the storm. Recent aircraft measurements show that Maxwell current output from thunderstorms is related to the lightning flash rate.

The ionospheric wind dynamo is produced in the region of the atmosphere where neutral winds can have the effect of moving an electrical conductor (the weakly ionized plasma) through the Earth's geomagnetic field. This produces an electromotive force that generates potential differences of 5 to 10 kV with a total current of 10^5 A extending over thousands of kilometers and flowing primarily on the dayside of the Earth.

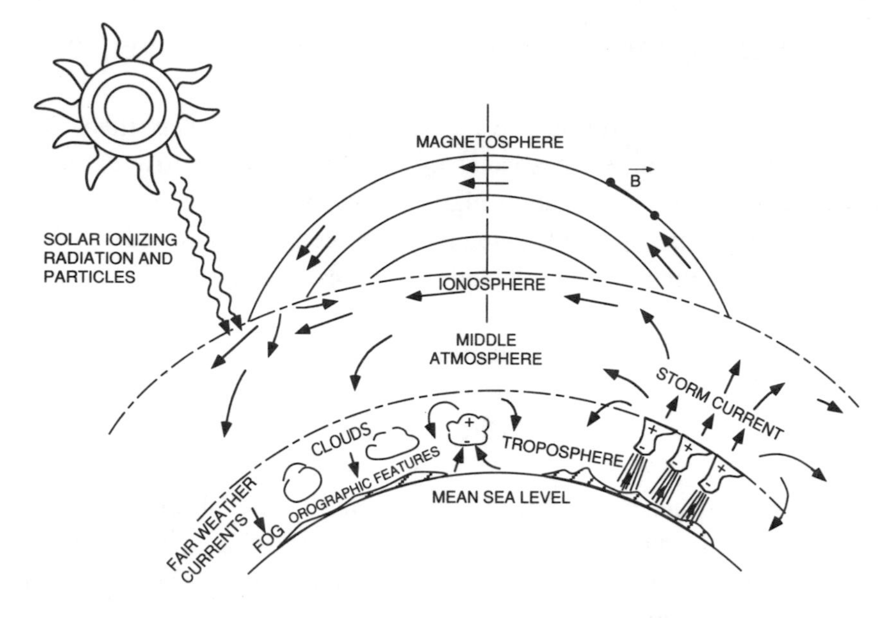

FIGURE 4.3 Schematic depiction of various electrical processes that make up the global electric circuit, illustrating how coupling between the different atmospheric regions connects the Earth's upper atmosphere to the biosphere. The ionosphere exists because of ionization by solar extreme ultraviolet radiation. From Studies in Geophysics: The Earth's Electical Environment, (NAS, 1986).

This current system is highly variable because of the changing tides and other disturbances propagating into the dynamo region from the middle atmosphere. Solar and auroral variability alter thermospheric winds and ionospheric electrical conductivities and thus influence the currents and fields in the dynamo region.

The flow of solar wind around and partly into the magnetosphere produces the solar wind/magnetosphere dynamo, which sets up plasma motion in the magnetosphere as well as producing electric fields and currents. This interaction is highly variable, but typically generates horizontal dawn-to-dusk potential drops of 40 to 150 kV across magnetic conjugate polar caps. The solar wind/magnetosphere dynamo is associated with a current flow of about 10^6 A between the magnetosphere and ionosphere. This generator depends on the properties of the solar wind flowing past the Earth's magnetosphere as well as on any internal current flows within the magnetosphere. The magnetosphere and the Earth's near-space environment are more fully discussed in Chapter 5.

There is considerable variability in tropospheric electrical parameters that influence the properties of the global circuit. These parameters include global cloudiness as it affects the electrical conductivity; turbulence in the planetary boundary layer; aerosols; pollution; radioactive ion production near the Earth's surface; fog; and surface processes in grasslands, forests, deserts, and ocean spray. Many of these processes have been studied in isolation, but their combined impact on the global electric circuit has not been properly evaluated.

Important processes in the middle atmosphere also have implications for variability in the global circuit. Aerosols produced by volcanoes can affect the electrical conductivities and electric fields. Rocket measurements of electric fields in the mesosphere indicate strong departures from Ohm's law, suggesting the presence of an as yet unidentified generator operating at mesospheric heights. Electrical conductivity enhancements in polar regions associated with energetic particle precipitation, solar proton events, and Forbush decreases in cosmic ray fluxes following solar eruptions all influence the properties of the global circuit in ways that are not well understood at present.

Electrical Couplings Between the Upper and Lower Atmospheres

Large horizontal electric fields (100 to 1000 km) generated within the ionosphere project downward in the direction of decreasing electrical conductivity, effectively down to the ground. Small horizontal electric fields (1 to 10 km) are rapidly damped as they map downward into the atmosphere from ionospheric heights. Since the electrical conductivity of the Earth's surface is large, horizontal electric fields cannot be maintained there, and a vertical electric field variation results to accommodate horizontal variations of ionospheric potential. Calculations show that the solar wind/magnetosphere generator can increase or decrease by up to 20 percent the air-Earth current and ground electric field at high latitudes during geomagnetic quiet times, with larger perturbations during geomagnetic storms. The magnitude of the ground variations also depends on the alignment of the potential pattern over the ground, being much enhanced by mountainous terrain in the polar region.

While electric and magnetic field coupling between the upper and lower atmospheres is well established physically, the impact on processes in the troposphere and biosphere and on other processes important for global change is unknown. Furthermore, human activities are slowly changing the atmospheric structure, atmospheric and ionospheric composition, aerosol loading, land surface properties, and other tropospheric variables, all of which will probably have some influence on the properties of the global electric circuit (Price and Rind, 1994).

SOLAR FORCING AND GLOBAL CHANGE WITHIN THE UPPER ATMOSPHERE

That the Earth's upper atmosphere is forced directly by variable solar energy inputs on all time scales is well established. Global mean temperature, global wind circulation, and constituent particle densities change continuously, in response to changing solar activity throughout the 11-year cycle (e.g., Evans, 1982), with the extent of the changes generally increasing with altitude. From the minimum to the maximum of the Sun's activity cycle, upper atmosphere temperatures increase by many hundreds of degrees (Figure 1.2), a direct consequence of solar EUV and UV heating. Greenhouse gases such as carbon dioxide and methane contribute to the radiative balance of the Earth's upper atmosphere as well as the lower atmosphere. Carbon dioxide cooling is the dominant cooling mechanism in the atmospheric region between 70 and 200 km; infrared cooling by CO_2 is largely responsible for the temperature minimum near 80 km shown in Figure 1.2.

Most studies of the climate change anticipated from the anthropogenic loading of greenhouse gases focus on the effects on the troposphere and middle atmosphere (Rind et al., 1990; Hansen et al., 1993). Recognition that trace gases released into the Earth's atmosphere from human activity could perturb the climate of the Earth provided much of the motivation for the USGCRP. As discussed in Chapter 2, these studies suggest that the troposphere will warm and the middle atmosphere will cool as trace gas concentrations increase into the twenty-first century. In the upper atmosphere, as in the middle and lower atmospheres, increases in anthropogenic gases are expected to affect the energy balance between solar heating and infrared cooling.

Recent studies have shown that trace greenhouse gases could effect considerable change in the structure of the Earth's upper atmosphere (Roble and Dickinson, 1989; Rishbeth, 1990). The global mean temperature of the upper atmosphere has been projected to cool by 10 K at altitudes near 70 km, and by 50 K around 150 km, in response to doublings of CO_2 and CH_4 concentrations (Figure 4.4). These changes will be superimposed on upper atmosphere temperature variations of many hundreds of degrees generated by changes in solar energy inputs throughout the 11-year activity cycle (Figure 1.2). Concomitant redistributions of major and minor constituents should occur throughout the entire atmospheric region. In the thermosphere, the atmospheric density at a given altitude has been projected to decrease by as much as 40 percent. The atmospheric scale heights that govern thermospheric and ionospheric properties should also be reduced, and the peak height of the ionospheric F2 region may be lowered by 20 km.

As a result of changes in the basic thermal and compositional structure of the atmosphere, increases in CO_2 may also damp the response of the thermosphere and ionosphere to solar and auroral variability. These changes also could affect the propagation of atmospheric tides, gravity waves, and planetary waves into the thermosphere. It is not clear how changes in the basic atmospheric structure and dynamics will affect the ionospheric wind dynamo and the coupling of the ionosphere and magnetosphere with the solar wind, but changes in thermospheric circulation might result in a changed electrodynamic structure of the upper atmosphere, and through dynamo action, alter magnetosphere/ionosphere coupling processes, as well as the entire terrestrial electric field system.

COUPLINGS OF THE UPPER ATMOSPHERE TO THE LOWER ATMOSPHERE

The most direct coupling between the upper and lower atmospheres is electrical. As discussed previously, large horizontal electric fields map almost unattenuated to the Earth's surface, where they perturb the vertical electric field maintained by global thunderstorm activity. In addition, the

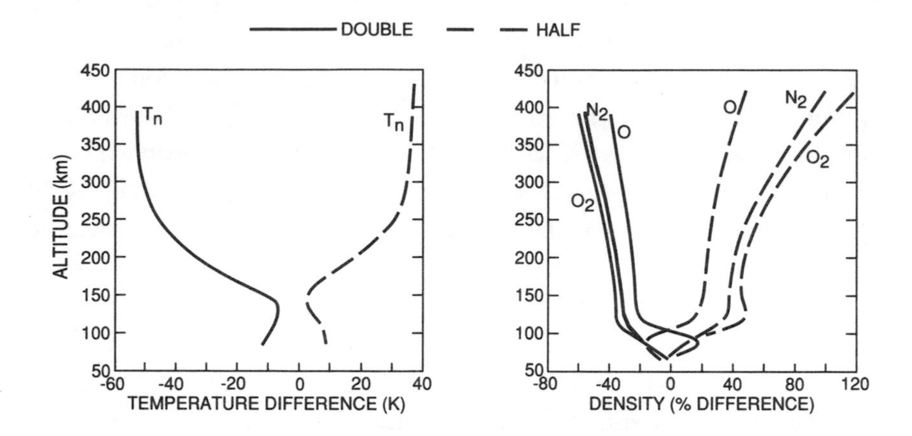

FIGURE 4.4 Temperature and density changes in the upper atmosphere as a function of the altitude in km predicted by the model calculations of Roble and Dickinson (1989) as a consequence of increasing greenhouse gases released in the lower atmosphere. Shown on the left are changes in the altitude profile of the neutral gas temperature for doubled and halved concentrations of CO_2 and CH_4. On the right are the corresponding percent changes in the density profiles of the primary upper atmosphere constituents. From R. Roble and R. Dickinson, Geophys, Res. Lett. 16, 1443, 1989, copyright by the American Geophysical Union.

geomagnetic quiet-time dynamo current and highly variable auroral current systems induce magnetic currents in the Earth's crust and alter the ground electric field. These perturbations have demonstrable impacts on human infrastructure, such as power networks and oil pipelines (Allen et al., 1989), but as yet unknown impacts on biological and atmospheric physical processes.

Also suspected are chemical and dynamical couplings between the upper and lower atmospheres. Nitric oxide is produced by solar soft X-ray and EUV radiation and auroral particle dissociation of molecular nitrogen into excited and ground states of atomic nitrogen that react with molecular oxygen. NO concentrations change significantly as a result of larger variations in solar soft X-ray fluxes (Siskind et al., 1990). Nitric oxide is important for understanding possible radiative and dynamical couplings of solar variability effects in the thermosphere with the middle atmosphere (Garcia et al., 1984; Siskind, 1994). For example, if transported downward to the region of 40 km, NO may have an important influence on the ozone density of the stratosphere (Huang and Brasseur, 1993). While the catalytic destruction of ozone in this region is relatively well understood (see Chapter 3), the transport of thermospheric-generated nitric oxide into the region is not. Because the photochemical destruction lifetime is on the order of one day throughout the sunlit mesosphere, NO must move downward during the polar night to reach the upper stratosphere. Two-dimensional modeling studies, as well as spacecraft observations of middle atmospheric ozone abundances, suggest that this may indeed be a viable coupling mechanism (Garcia et al., 1984; Solomon and Garcia, 1984). Changes in nitrate content of antarctic snow associated with solar activity may reflect these processes (Dreschhoff and Zeller, 1990).

In addition to solar-related processes, the lowest layers of the upper atmosphere are influenced by turbulent breaking of atmospheric gravity waves and tides and by sporadic, intermittent compositional exchanges of atomic oxygen and nitric oxide. These complex turbulent exchange processes influence long-lived species such as carbon dioxide, carbon monoxide, water vapor, and atomic and molecular hydrogen. Although theoretical studies have indicated that chemical, dynamical, and radiative interactions are a viable coupling mechanism, the whole question of thermospheric/lower atmospheric exchange is not well understood, primarily because our atmosphere between about 50 and 200 km is virtually unexplored on a global basis.

5

Solar Variations and the Earth's Near-Space Environment

BACKGROUND

The geospace environment is the region of transition between the Earth's protective atmosphere and the onrushing solar coronal plasma (the solar wind). It is a region of even greater variability than the Earth's upper atmosphere, transmitting and redepositing the fluxes of mass, momentum, and energy received from both the Sun and the Earth. The near-space environment responds dramatically to changing solar energy inputs (e.g., Gorney, 1990), but this solar forcing appears to have little direct impact on the Earth's climate.

The Earth's near-space environment does provide a critical buffer between the highly dynamic space environment and the relatively placid lower and middle atmospheres. To a large extent, it determines the penetration of these layers by energetic particles accelerated on the Sun, from outside of the solar system, and within the magnetosphere. As discussed in Chapter 3, both energetic solar protons and relativistic electrons can destroy ozone and affect the middle atmosphere. Also, large ejections of mass and magnetic fields from the Sun, whose influence is transmitted to the Earth through the near-space environment, can significantly affect certain complex technological systems, including electrical power grids, Earth-orbiting spacecraft, and communication links (Joselyn, 1990).

THE SOLAR WIND AND THE EARTH'S MAGNETOSPHERE

Flowing from the Sun is the solar wind, which continuously carries magnetized plasma and energetic solar particles into the vicinity of the Earth. The Earth and its atmosphere are shielded from the direct impact of these particles and plasmas by the magnetosphere, a relatively self-contained region in space whose global topology is organized by the intrinsic magnetic field of the Earth. This field, which may be represented to a reasonable approximation by a dipole originating in the Earth's molten metal core, extends far into space and serves to deflect the onrushing solar wind. The stand-off distance (the magnetopause), commonly about 10 Earth radii (R_E) at the subsolar point, depends on the solar wind pressure and is highly variable. In the outer reaches of the Earth's near-space environment, tangential stresses applied by the solar wind set up a system of boundary region currents that effectively constrain the outer geomagnetic field to a comet-shaped form with a long tail extending downstream from the Sun (Figure 5.1). Thus, the Earth's magnetosphere extends from the upper atmosphere/ionosphere to altitudes of about 10 R_E on the sunlit dayside and to more than 1000 R_E on the nightside.

Mass, momentum, and energy are imparted to the magnetosphere with great variability by the continuously flowing solar wind. The primary form of plasma energy available at 1 astronomical unit (AU) is kinetic, as a result of the motion of the solar wind relative to the Earth. Solar wind plasma interacts with the projected cross-section of the entire magnetosphere (a disk of radius about 20 R_E), so that the total power intercepted due to the solar wind kinetic energy is about one thousandth of the radiant energy intercepted by the disk of the Earth. This energy transfer occurs with much greater variability than the radiant heating variations associated with the 0.1 percent solar cycle change in total solar irradiance. However, it is not the solar wind kinetic energy flux *per se* that seems to control geomagnetic activity, but rather the embedded solar wind magnetic field.

The major processes that extract, store, and dissipate energy from the solar wind flowing past the Earth, subsequently disturbing the geospace environment, involve the generation of plasma and energetic particles from stored magnetic fields. Three primary forms of energy dissipation detectable in the Earth's atmosphere are auroral particle precipitation,

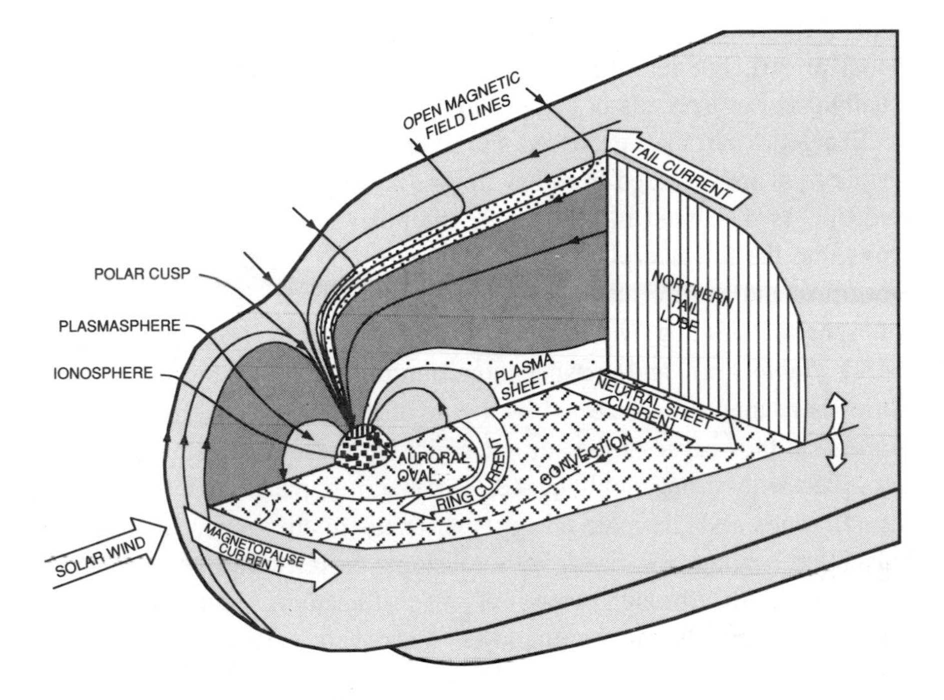

FIGURE 5.1 The Earth and its atmosphere are surrounded by the near-space environment. The solar wind carries magnetized plasma and energetic solar particles into the vicinity of the Earth, which is shielded from their direct impact by the magnetosphere, a relatively self-contained region in space whose global topology is organized by the magnetic field associated with the Earth. Courtesy of T. Potemra, NASA publication.

auroral Joule heating, and energetic neutral atoms produced from extraterrestrial ring current flows. Eventually, plasma particles convert part of their energy to radiation modes such as auroral displays and kilometric radiation.

SOLAR ERUPTIVE EVENTS AND GEOMAGNETIC STORMS

Explosive outbursts from the Sun release energy primarily in the form of X-rays, UV radiation, energetic particles, magnetized plasma, and shock waves. Large injections into the magnetosphere of magnetized plasma from the Sun generate major disturbances called geomagnetic storms. Moderate magnetic storms may occur relatively frequently (every month or so), but really large storms due to major solar disturbances usually occur at intervals of many years.

Energetic particles produced by solar eruptions (and also galactic cosmic rays) are excluded from low magnetic latitudes by the geomagnetic field but, as discussed in Chapter 3, the polar regions of the Earth are exposed to the full interplanetary flux (Figure 3.3). During geomagnetic substorms and storms, energized particles bombard the Earth's upper atmosphere, colliding with atmospheric constituents, transferring their energy (Table 1.1), and causing large auroral displays. Substorms and storms have long been detected by virtue of the intense magnetic disturbances that they cause in the auroral regions. These magnetic effects are associated with strong field-aligned (Birkeland) currents that flow in the auroral zones and dissipate energy resistively in the upper ionosphere. This Joule heating associated with substorm currents can be monitored from the Earth (through arrays of magnetometers), and ionospheric conductivity models can be employed to convert measured currents to ohmic dissipation.

One of the primary manifestations of a geomagnetic storm is a large enhancement of the extraterrestrial ring current, composed of trapped particles drifting in the Earth's inner magnetospheric region. During such enhancements the ring current can cause large magnetic disturbances in the low-latitude magnetic field at the Earth's surface. Accelerated particles and plasma are injected from the tail of the magnetosphere into the ring current. There, these particles gradually lose their energy (over hours or days) due to precipitation and charge-exchange processes. Hence, the ring current is a major sink of magnetospheric energy.

A significant part of the energy dissipated during geomagnetic activity can be assessed by examination of auroral and ring current terms. Over the years, indices of auroral disturbance (e.g., the AE index) and ring current disturbances (e.g., the Dst index) have been formulated. These are basically parameters of levels of magnetic disturbance as measured on the Earth's surface, and they calibrate the disturbance level. In turn, it has been possible to assess magnetospheric energy losses in terms of these indices.

TERRESTRIAL IMPACTS

Energetic solar outbursts impact the Earth in a variety of ways, depending on how the released energy couples into the global system through the Earth's near-space environment. The impact of solar outbursts can be severe. For example, associated with the March 1989 eruption were high-frequency communication outages and disruptions of navigation, anomalous radar echoes at high latitudes, and electrical power outages including a 9-gigawatt failure in Quebec that affected 6 million people for half a day and caused millions of dollars of damages. Heating of the Earth's atmosphere during the storm increased satellite drag, leading to uncontrolled tumbling of several satellites and more rapid orbital decay of others (Allen et al., 1989; Joselyn, 1990). Depletion of stratospheric ozone over Antarctica may also have occurred (Stephenson and Scourfield, 1991). Particle events such as these could create serious radiation hazards to manned space missions at high orbital inclination or outside the magnetosphere.

Huge disturbances in the polar ionosphere often result from geomagnetic storms, and intense auroral luminosity can reach over much of the high-latitude portion of the Earth. Low-latitude magnetometers on the Earth's surface can register changes up to about 1 percent of the normal ambient field. Such changes in the magnetic field can have significant short term effects on navigation, resource exploration, and other human activities. Particle events associated with eruptions on the Sun cause failures in microelectronic circuits, buildup of electric charge on spacecraft, and potentially harmful radiation doses to crews of high-altitude aircraft (Joselyn and Whipple, 1990). Clearly, the highly sporadic and unpredictable nature of geomagnetic activity makes it very difficult to estimate the importance of its effects on the terrestrial environment on the time scale of decades to centuries. In an historical context, knowledge of the role of the geomagnetic field itself is important because variations in its strength modulate the exposure of the Earth's atmosphere to bombardment by galactic cosmic rays, allowing variations in production of ^{14}C and other cosmogenic nuclides that could mimic variations in solar activity.

Longer term terrestrial influences may also arise from the fluxes of relativistic electrons with energies of several million electron volts, frequently present within the magnetosphere, that can reach to depths similar to those reached by solar protons. These energetic electrons are accelerated primarily within the magnetosphere (rather than on the Sun) and occur much more frequently than major flare events. The presence or absence of the relativistic electron fluxes appears to be strongly controlled by the existence of high-speed solar-wind streams emanating from persistent coronal holes, which in turn are strongly controlled by the solar activity cycle. The extent to which precipitating relativistic electrons actually enter the atmosphere is currently uncertain. As noted in Chapter 3, if a significant amount of these ionizing particles reaches the middle atmosphere, the long term effects on ozone concentrations in the stratosphere could be significant (Callis et al., 1991).

6

Understanding the Variable Sun

BACKGROUND

Changes in the energy from the Sun potentially could influence global change directly by modifying the Earth's surface temperature (Chapter 2) and by creating and destroying atmosphere ozone at variable rates (Chapter 3). Solar variability may also influence global change indirectly, by modifying the middle atmosphere, which is connected chemically, dynamically, and radiatively with the troposphere/biosphere (Chapter 3). In the upper layers of the Earth's atmosphere, and in the geospace environment, solar variations cause dramatic changes that are critical for understanding the processes within those regions, although the extent to which these changes couple to lower atmospheric layers (Chapters 4 and 5) is uncertain.

Observations over the past decade have provided an exciting perspective on how the Sun's energy inputs to the Earth change with time. In this period were obtained the first long term records from space of the solar radiative energy inputs to the Earth that are critical for studying solar influences on global change: total solar irradiance and the solar UV spectral irradiance, as well as fluxes of energetic protons and electrons. Ground based measurements were also made of solar observables closely related to the energy inputs measured from space. Physical associations between open field regions on the Sun, high speed solar wind streams, coronal mass ejections, and geomagnetic activity were established through a variety of space missions. Taken together, these observations have revealed new insights into how solar magnetic activity modulates terrestrial solar energy inputs and how magnetized plasma from the Sun evolves as its flows to the Earth. These observations have established beyond doubt that the Sun's energy output varies continuously on all observed time scales.

Predicting, understanding, and monitoring global change are the ultimate objectives of the USGCRP (Chapter 1). Yet contemporary measurements of solar energy inputs alone reveal little about future solar variability nor of past solar variations that might have influenced the paleoclimate record, which is the focus of the Earth System History science element of the USGCRP. To begin to understand how the Sun varied in the past and how it might vary in the future, we must first understand why the Sun varies at all.

The fundamental physical processes that generate the variations observed in solar energy production are associated with the 22-year magnetic cycle of the Sun. The sunspot number time series remains the principal historical indicator of this cycle, and it is shown in Figure 6.1. This is the record of solar activity that was compared with the ^{14}C and temperature time series in Figure 1.3 and with surface temperature anomalies in Figure 2.4. Recent monitoring from space indicates that both the total solar irradiance (Figure 2.1) and the UV irradiances (Figure 3.2) increase near the peak of the sunspot cycle and decrease during times of few sunspots. Likewise, the flow of energy, plasma, and magnetic fields from the Sun into the Earth's environment depends on the magnetic cycle. Fundamental to understanding the Sun's behavior as a variable star is understanding how variations in its emitted energy are generated from the magnetic activity cycle.

ORIGINS OF SOLAR VARIABILITY

The 22-year magnetic cycle of the Sun manifests itself as the familiar 11-year sunspot cycle, the 22-year cycle being simply two 11-year cycles having reversed magnetic field polarities. Physically, the sunspot cycle is a roughly periodic emergence, approximately every 11.1 years, of strong magnetic flux tubes at the solar surface in the form of sunspots. More

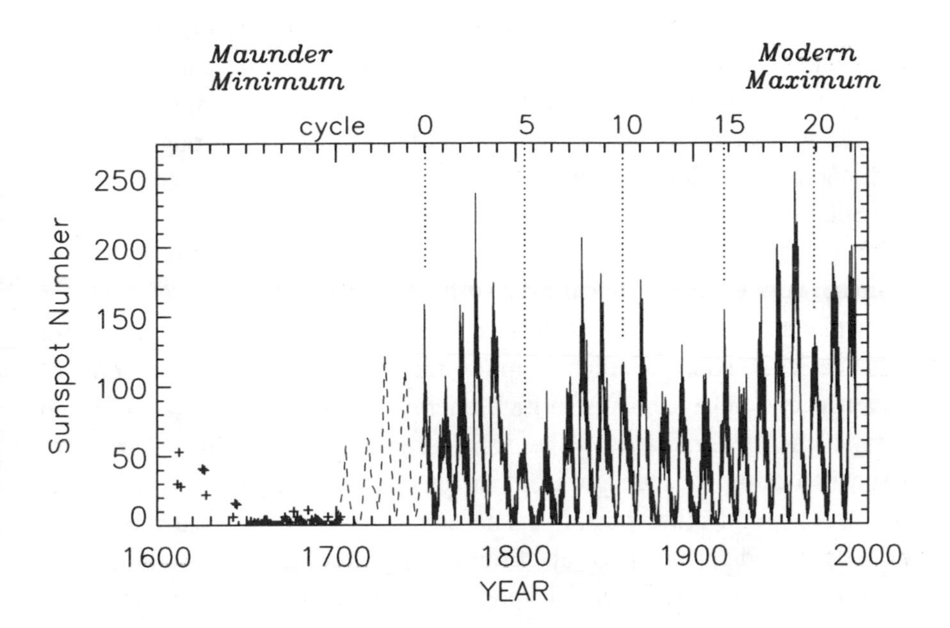

FIGURE 6.1 Solar activity variations during the past four centuries, as indicated by monthly means since 1750 of the sunspot number (solid line) with yearly means from 1610 to 1750 (from the National Geophysical Data Center, dashed line, and Eddy, 1976, crosses). Prominent in the record is a cycle of about 11 years. According to Eddy, the 11-year sunspot cycle was severely depressed between 1645 and 1715 (see also Ribes et al., 1989). Thus, during the contemporary epoch when the Sun has been observed most intensely, its activity has been at relatively high levels compared with the past 300 years. This is confirmed by cosmogenic isotope records (McHargue and Damon, 1991). The Sun is likewise at high activity levels compared with other Sun-like stars (Figure 2.3). Recent analysis of sunspot observations in the 18th century indicate that the sunspot number may overestimate solar activity levels in that period (Hoyt et al., 1994). From J. Lean, Reviews of Geophysics, 29, 506, 1991, copyright by the American Geophysical Union.

generally, the solar activity cycle pertains to the periodic emergence of magnetic flux that generates not just sunspots, which are dark, but a variety of phenomena, especially bright regions known as plages and faculae that radiate strongly at UV and EUV wavelengths. The dark sunspots and bright plages and faculae modify the radiation from the solar disk, thereby generating the variations observed by spaceborne solar radiometers. Also, changes in the Sun's magnetic field topology, due to both flux tube emergence and latitudinally differential rotation of the solar atmosphere, generate field configurations that lead to transients such as solar flares and coronal mass ejections, and longer lived features such as coronal holes. These latter phenomena affect the Earth through input of high energy particles and plasma into the geospace environment.

Solar flares and solar global oscillations are prominent examples of solar energy variations on time scales of seconds to hours. Flares, although very energetic, occur over sufficiently small fractions of the solar hemisphere that even the largest and most energetic of them do not enhance total solar irradiance more than a hundredth of a percent. Nevertheless, enhancements in high energy emissions, such as EUV and X-rays, can be dramatic. The solar eruptions with which flares are associated also produce significant fluxes of energetic particles (electrons and protons and other nuclei). The enhanced EUV and particle fluxes from flares and coronal mass ejections can significantly alter the ionization state of the Earth's middle and upper atmosphere. At times when many eruptions occur in succession, the effects may persist for many months, as demonstrated by the semicontinuous solar proton events during 1989-1990 (Reid et al., 1991), discussed in Chapter 3.

Like flares, global oscillations of the Sun, such as the five minute p-mode oscillations, have minimal effect on total solar irradiance -- on the order of 3 parts per million of the total flux (e.g., Hudson, 1987). The significance of the solar oscillations is that they provide a unique technique for sensing physical properties and variations of the solar interior (Leibacher et al., 1985; Gough and Toomre, 1991), thus providing crucial insight into the mechanisms of the solar cycle.

The Sun's 27-day rotation modulates the steady radiative and plasma outputs from the Sun because at times (especially near activity maxima) the sources of these outputs are localized in narrow heliocentric longitudinal bands and are sufficiently long-lived that they reappear on successive rotations (see Lean, 1987, for examples). Rotational modulation is clearly seen in the total solar irradiance, with a rather complex waveform that reflects the competing effects of dark spots and bright faculae. Although 27-day periodicity is attributable to the Sun's rotation, no physical models explain the emergence, evolution, and decay of the magnetic active regions on the solar surface that cause changes in the amplitude and phase of this cycle. Since the 27-day rotation modulates solar irradiance by changing the amount of active region emission seen at the Earth, observations and investigations of this periodicity aid in understanding the origins of the geomagnetic and photon flux variations that are important for the terrestrial environment (e.g., Lean, 1987).

Nor is the emergence of magnetic activity with an 11-year periodicity properly understood. Conceptually, a dynamo process is thought to underlie the Sun's magnetic cycle (e.g., DeLuca and Gilman, 1991), and models have been constructed to provide insight into the specific basic solar characteristics that must change to produce the periodic magnetic flux tube emergence that is responsible for solar energy input variations to the Earth. These models are necessarily theoretical, based on the internal interactions of a plasma rotating differentially within a convective envelope. No widely accepted solar dynamo model can reproduce all of the observed features of the solar cycle. A particular problem is the models' inconsistency with helioseismologic observations of the internal rotation of the Sun (e.g., Leibacher et al., 1985), an essential component of any model of flux emergence.

Even though the 11-year solar activity cycle appears to be the fundamental cause of changes in terrestrially sensed energy on decadal time scales, the possibility that long term changes may also be occurring in the Sun because of non magnetic mechanisms cannot be ignored. Short (five-minute) p-mode oscillations arising from a natural cavity resonance inside the Sun are well established. These oscillations are a property of the global Sun, a star, as opposed to localized sources of fluctuation such as sunspots and plages/faculae. Other solar features, such as ephemeral magnetic regions and the chromospheric grains seen in Ca II K, contribute to the energy output and are widely distributed over the Sun.

Over times much longer than the 11-year activity cycle, there is the possibility of superimposed slow secular change in the Sun's energy inputs to the Earth. The Maunder Minimum (Eddy, 1976), seen clearly in the sunspot record in Figure 6.1, is evidence for such change and the only example in the contemporary solar record. The absence of sunspots and possible supression of the 11-year cycle during this extended period suggests that the flux emergence process may have stopped completely for several decades. Relative to the present, the Sun's internal circulation may have been substantially different (Eddy et al., 1976), and its diameter larger (Nesme-Ribes et al., 1993). Indirect evidence for many similar episodes in solar behavior comes from the radiocarbon and auroral records (Eddy, 1976), but only very recently has it been recognized that substantial decreases in the Sun's radiative output might have accompanied these episodes (White et al., 1992; Lean et al., 1992a). The ^{14}C data also suggest that solar activity was very high in the twelfth century, an epoch corresponding to the Medieval Warm Period of approximately 300 years (Figure 1.3). Eddy (1976, 1977) suggests that the total irradiance may follow the envelope of the sunspot cycle curve, waxing and waning on century time scales. These findings indeed suggest a relationship between the Sun and the climate, but a solar variability model that describes the Sun's energy inputs to Earth at times in the past has yet to be developed.

Because the Sun is apparently a normal star, insights into solar variability can be gleaned from observations of Sun-like phenomena in stars with mass, age, and rotation rates similar to the Sun's (Baliunas, 1991). In particular, comparative solar and stellar Ca II emission measurements indicate that the activity levels of the contemporary Sun correspond to the highest levels observed in other stars (White et al., 1992). These observations also suggest that times of arrested activity, as exemplified on the Sun by the Maunder Minimum, may be common in other stars, and that such times appear to be accompanied by reduced energy output (Baliunas and Jastrow, 1990; Lean et al., 1992a).

RELATIONSHIP BETWEEN SOLAR SURFACE STRUCTURE AND ENERGY FROM THE SUN-AS-A-STAR

To better understand the relationships between changes on the Sun and changes in the Earth's atmosphere, it is important to understand how measurements made at the Earth at 1 astronomical unit (AU) project back to structures on the solar surface.

Radiation

The transmission path of solar radiative input to the Earth is line-of-sight. Active regions (sunspots, plages and faculae, filaments, coronal holes, etc) modify the local intensity of the solar surface, resulting in an inhomogeneous solar disk whose radiation field is variable in both time and direction. As shown in Figure 6.2, these surface features evolve continuously throughout the solar activity cycle, and they have the largest effect near times of activity maxima. The radiative energy input to the Earth is the integral of the radiance from the entire solar hemisphere visible at the Earth -- that is, the irradiance. As a result of this integration, the interpretation of variations observed in the total and spectral solar irradiances involves constructing irradiance time series by combining the contributions from sunspots, plages, the bright magnetic network, and internetwork regions. In this way, variations in both total and ultraviolet spectral irradiance can be traced to the changing populations of active regions on the solar disk.

Ground based observatories measure the position, brightness, and area of sunspots, plages, and faculae daily (e.g., Beck and Chapman, 1993). These data have been combined in simple empirical models to reconstruct the measured irradiances (Cook et al., 1980; Lean et al., 1982; Foukal and Lean, 1988; Willson and Hudson, 1991). Comparisons of the model calculations with the measured irradiances show that the principal contributors to radiative variability are sunspots, plages/faculae, and chromospheric network, both on solar rotation time scales as well as over the decadal scale of the solar cycle. Residuals between the measurements and reconstructions are analyzed for the possibility that they arise from experimental error, incorrect assumptions in the models, or missing energy in storage deep in the Sun.

In making the connection between fluctuations in irradiances and solar surface inhomogeneities, two types of ground based data have been crucial. One is photometric data on sunspots, faculae, and plages, and the other is spectral irradiance data; both are measured in appropriate regions of the solar spectrum. The photometric data provide records of the individual active regions that generate irradiance variations (Figure 6.2). Sunspots are identified most clearly in white-light solar spectroheliograms, whereas images in the Ca II K line remain the principal source of photometric data on plages and the network, features that are also seen clearly in solar images in other spectral lines, such as the He I 1083 nm line. The spectral irradiance (Sun-as-a-star) data measure the integrated effects of these active regions on layers of the Sun from which the total and UV radiations are also emitted. Figure 6.3 shows three of the most informative solar records: the Sun-as-a-star chromospheric Ca II and He I indices (White

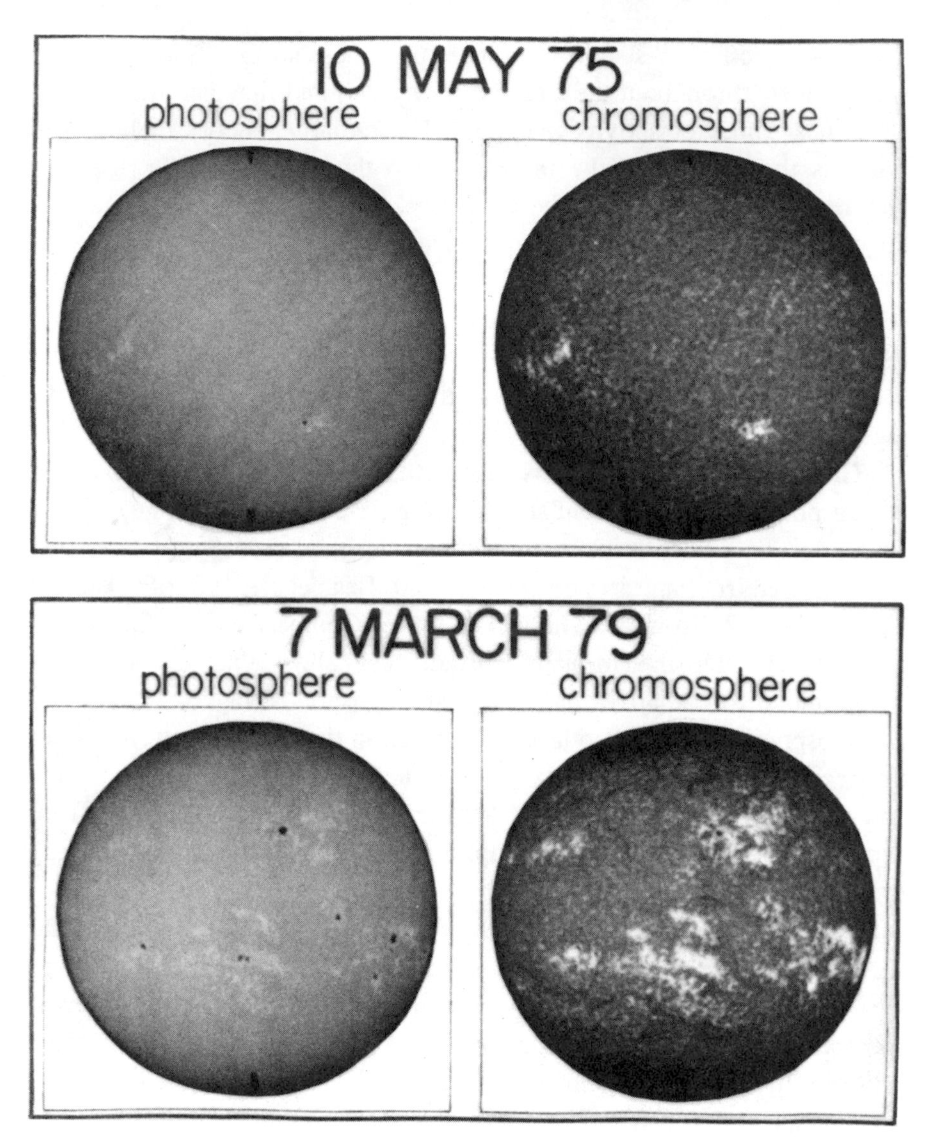

FIGURE 6.2 Active regions on the Sun are seen in spectroheliograms of the photosphere and chromosphere, taken in the wings and at the center, respectively, of the Ca II K line during solar minimum (1975) and solar maximum (1979). Surface inhomogeneities associated with magnetic active regions can been seen as regions of enhanced brightness (plage and faculae) and also as small dark regions (sunspots). Both the area and the number of such magnetic regions increase at times of solar activity maxima, relative to activity minima. Spectroheliograms from R. Howard. From J. Lean, J. Geophys. Res., 92, 843, 1987, copyright by the American Geophysical Union.

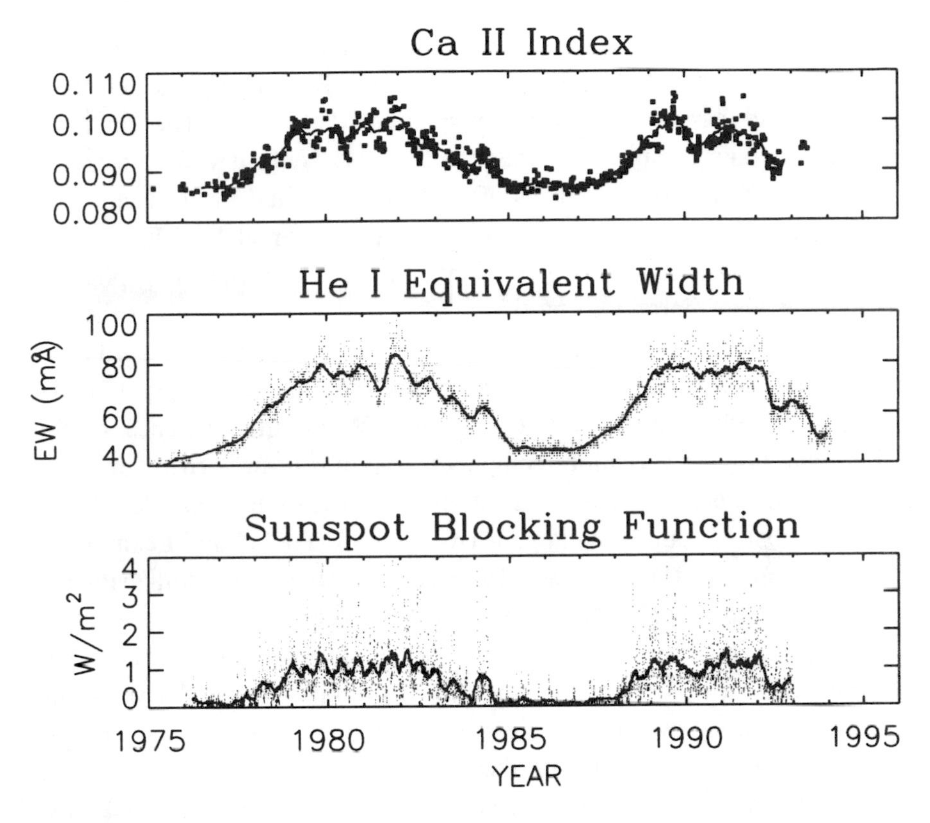

FIGURE 6.3 Variations during solar cycles 21 and 22 of ground-based solar indices of high relevance for interpreting solar irradiance variations. In the upper panel is the chromospheric Ca II K index (ratio of the core to wing emission of the Fraunhofer line at 393.37 nm) (White et al., 1990). In the middle panel is the primarily chromospheric He I 1083 nm line equivalent width (Harvey, 1984; Harvey and Livingston, 1994) and in the lower panel is the sunspot blocking function (calculated according to Foukal, 1981). In each figure, the solid lines are the variations smoothed over approximately three 27-day solar rotations. Both the Ca II and He I time series are useful proxies for solar radiative output from bright faculae while the sunspot blocking is a parameterization of the emission deficit caused by dark sunspots. Based on J. Lean, Reviews of Geophysics, 29, 513, 1991, copyright by the American Geophysical Union.

and Livingston, 1981; Harvey, 1984; Livingston et al., 1988; White et al., 1990) and the sunspot blocking. In this regard, the solar 10.7 cm radio flux is also important, especially in an historical context, since it was the only Sun-as-a-star indicator measured during solar cycles 19 and 20.

Figure 6.4 shows a comparison of the total solar irradiance measured by ACRIM I (on SMM) and by Nimbus 7/ERB with models developed from ground based data for the respective data sets. The rotational modulation data during 1982 illustrate that the day-to-day variations that arise from the competing effects of dark sunspots and bright faculae are well reproduced by the Foukal and Lean (1990) model. Much of the longer term variability during solar cycle 21 and the ascending phase of cycle 22 is also reproduced by this model, with the exception of the first years of the record. The longer term solar cycle changes occur because of a brightness component in addition to the sunspots and the brightest faculae associated with magnetic active regions. While the existence of this additional 11-year variability component has not been verified by direct observation, it is thought to reside, at least in part, in the network of bright emission that surrounds the large active regions (Foukal and Lean, 1988; Foukal et al., 1991). It may also have a global (i.e., non magnetic) component (Kuhn et al., 1988; Kuhn and Libbrecht, 1991). Discrepancies between the measurements and the model during 1979-1980 may be instrumental in origin. If not, they raise the possibility of a variability component acting over time scales longer than the 11-year cycle (Lee III et al., 1994). The utility of these models, and the need to resolve discrepancies with the measurements, emphasize the need for high precision image data in the interpretation of irradiance time series in the future.

Statistical comparisons indicate that the Ca II and He I indices provide reconstructions of solar UV irradiance variations superior to those afforded by the 10.7 cm radio flux, over both solar rotation and solar cycle time scales (Barth et al., 1990; Bachmann and White, 1994). Nevertheless, the 10.7 cm flux has been used extensively for the past few decades as a proxy for EUV and UV irradiance variations in terrestrial applications. For example, essentially all analyses of ozone data for the purpose of extracting long term trends have used the 10.7 cm flux, in lieu of UV irradiance data, to account for solar forcing of ozone changes (Stolarski et al., 1991; Hood and McCormack, 1992; Randel and Cobb, 1994; Reinsel

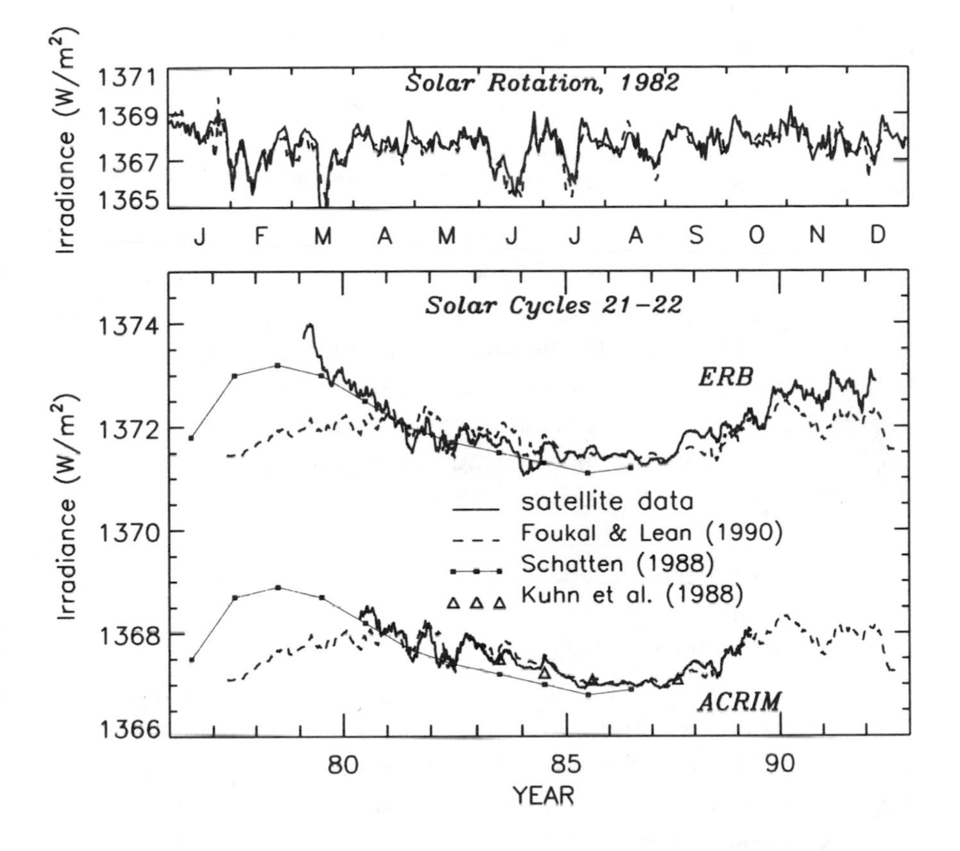

FIGURE 6.4 Comparison of the SMM/ACRIM I and Nimbus 7/ERB total solar irradiance data with empirical variability models constructed from the respective radiometry. In the upper panel, daily values of a model developed by Foukal and Lean (1990) show good agreement with the measurements over solar rotation time scales. In the lower panel 81-day running means during cycle 21 and cycle 22 of the solar irradiance data and the Foukal and Lean (1990) model show similar variations over time scales of active region evolution. Differences between the measurements and models are largest near maximum solar activity, especially prior to 1981. In Schatten's (1988) model, maximum total irradiance is postulated to occur prior to the peak of the activity cycle because of the role played by faculae at high heliocentric latitudes, which are thought to occur more frequently during times of minimum solar activity. This high latitude facular emission was postulated to help explain the discrepancy, possibly of instrumental origin, between the Foukal and Lean model and the irrandiance data prior to 1981. Based on J. Lean, Reviews of Geophysics, 29, 520, 1991, copyright by the American Geophysical Union.

et al., 1994b) (Chapter 3). With the availability of ground based data that better reflect the processes that generate solar radiative output variations, empirical models that predict these variations can be improved. Knowledge of the relationship between the 10.7 cm radio flux and solar radiative output nevertheless remains important for its historical relevance.

Plasma and Particles

Plasma from the solar corona flows radially outward from the Sun and impacts the Earth as the solar wind. Frozen into this plasma is the interplanetary magnetic field (the extended field of the Sun). Solar rotation, combined with the outward motion of the highly conducting plasma, winds this field into a spiral pattern and produces high speed and low speed streams. High speed solar wind streams (velocities of 700 to 850 km/sec) originate in regions where the magnetic field lines are open and connect to the interplanetary field; in contrast, the slow streams (velocities of 300 to 400 km/sec) come from regions above sunspot complexes where the magnetic field is closed to the solar surface. Shocks are formed when a high speed stream overtakes a low speed stream. Coronal mass ejections periodically disrupt the quasistationary pattern of high and low speed flows from coronal holes and streamers. The projection of solar wind profiles at the Earth back to their origins on the Sun can therefore be ambiguous, which emphasizes, again, the importance of knowledge of solar magnetic field structure in determining the time profile of solar energy input at 1 AU.

Solar energetic particles carried to the Earth by the solar wind interact with the Earth's magnetic field in ways that depend on the spectral distribution of their energy. The particles spiral along terrestrial magnetic field lines, entering the atmosphere primarily in the auroral zones surrounding the two geomagnetic poles (Figure 3.3). Projection of this complex field interaction backward to the Sun is aided by solar images in the visible spectrum that locate the sites of flares that are frequently associated with the eruptive events that are the sources of these particles. At times of high solar activity, however, the occurrence of numerous eruptions makes identification of the source of the particle flux more difficult.

Cosmic Rays

The diffuse cosmic ray flux that produces ^{14}C in the Earth's atmosphere (see Chapters 3 and 5) is modulated by solar activity on passing through the heliosphere (e.g., Lopate and Simpson, 1991). The particle flux at the Earth is highest when solar activity is at a minimum, so that the naturally archived ^{14}C record contains the signature of past solar variability excursions. Interpreting the ^{14}C record depends on understanding this modulation process as well as other effects, such as changes in magnetic field strength and in the terrestrial carbon cycle (Stuiver and Braziunas, 1993). Until there is a solid appreciation of the connection between the cosmogenic isotope variations and the modulation of the energy output from the Sun by magnetic active region phenomena (sunspots, plages, network), it will be difficult to construct reliable quantitative estimates of the strengths of extrema in solar energy, such as during the Maunder Minimum type episodes that appear to have occurred commonly (every 200 years or so) in the past.

REQUIREMENTS FOR IMPROVED UNDERSTANDING

Present

Since relatively reliable knowledge of solar radiative output variations exists for only the most recent 11-year solar cycle, much will be learned by the continuation and improvement of the total and UV spectral irradiance observations from space and of the ground based indicators of the various solar processes that cause and reflect their variations. The required ground based data are primarily white light images of sunspots and the Ca II and He I indices and images. The images (preferably of a few arc-seconds resolution or better, with accurate photometric calibration) are necessary not only for the formal construction of irradiance time series from ground based surrogates, but also to allow fundamental descriptions of spots, plages, faculae, and network as they evolve in time. Continuous daily measurements are needed at least for the duration of the Sun's magnetic cycle of 22 years, and more ideally for 100 years or longer, since the 88-year Gleissberg cycle may represent the dominant forcing factor.

The basic process that leads to variations in solar energy input to the Earth is the emergence and evolution of magnetic flux tubes in the solar atmosphere. The principal observational needs are for better photometric and positional data pertaining to the flux tubes that form sunspots, plages, faculae, the magnetic network and their motions on the solar disk. Theoretical models are needed to understand the basic energetics in these flux tubes and how the magnetic field interacts with solar convection to change the energy transport. Since this affects the brightness of magnetic structures throughout the solar spectrum, it bears directly on interpretation of the full disk observations.

An important tool in interpreting measurements of solar radiative output is the comparison of the measurements with synthesized solar spectra (Kurucz, 1991; Mitchell and Livingston, 1991; Avrett, 1991). Theoretical reconstructions of the solar spectrum, although based on one-dimensional thermodynamic and structure models, include millions of solar spectral lines from EUV to infrared wavelengths and provide insight into the spectral composition of the solar irradiance. At present it is not certain how the spectral energy distribution of the total irradiance changes over the solar cycle. The theory of the formation of the solar spectrum also establishes physical connections between different wavelengths, providing insight into the extent that variations at one wavelength mimic variations at another. Consideration of such connections may lead to more efficient observational programs in the future.

Magnetic flux tube emergence is thought to be associated with a dynamo lying deep inside the Sun. The goal of stellar dynamo models is to reproduce the periodic variation of flux tube emergence seen on the Sun and also to show the systematic latitude and polarity variations that occur as spots move from high latitudes toward the equator as the cycle progresses, with the polarity of sunspot pairs and the polar fields reversing from one cycle to the next. The necessary empirical boundary conditions for such models come from detailed knowledge of magnetic flux tubes, their distribution in both time and position on the Sun, and evolution of their energy transport seen in observations of radiative output. The most important constraints lie in the differential rotation structure of the solar interior and the properties of the interface between the convection zone and the Sun's radiative core, where the solar dynamo is thought to lie. Helioseismic observations continue to play a crucial role in studying the dynamo because they provide a way to probe the solar interior, but they will illuminate the dynamo problem only if measurements extend over solar cycle time scales.

Past

In the context of global change, understanding the past behavior of the Sun may well be essential for unraveling the paleoclimate record. Coincidences between climate change in the twelfth and seventeenth centuries and changes in ^{14}C as a proxy for solar activity (corroborated by the ^{10}Be record) suggest that there may be threshold levels of high and low solar activity at which the Sun begins to play a significant role in changing global climate (see Chapter 2). Knowledge of physical conditions on the Sun at these extrema is primitive because modern scientific observations have all been made during an era of high solar activity. Galileo's and others' discovery of sunspots at the beginning of the seventeenth century came at a time to document an ensuing period of low solar activity. Without this documentation, the possible role of the Sun in the Little Ice Age might still not be appreciated. Given the meager and often discontinuous evidence for solar variability in the past, inference of the physical state of the Sun at times of extrema is speculative but necessary. The first steps can be taken with empirical models now available (e.g., Foukal and Lean, 1990; White et al., 1992; Lean et al., 1992a; Hoyt and Schatten, 1993; Nesme-Ribes et al., 1993), but the credibility of such research would be strengthened immeasurably through development of a successful physical model of the solar cycle.

An expanded view of solar variability is provided by knowledge of cyclic behavior in other stars, afforded by long term measurements of stellar Ca II emissions in Sun-like stars. Baliunas and Jastrow (1990) present stellar cycle data that may indicate the presence of Maunder Minimum-type episodes in one-third of the observed stars, but their sample (13 stars) is so small that this conclusion must be regarded as speculative at this time. It is, however, consistent with an independent result from analysis of the radiocarbon data by Damon (1977). Furthermore, the distribution of Ca II emission exhibited by the Sun-like stars does appear to be consistent with the range of Ca II K emission seen in the present-day Sun (White et al., 1992). By assuming that during the Maunder Minimum, the mix of active region emission that generates the solar Ca II irradiance was somewhat different than is currently seen in the Sun, the radiative output of the Sun can be estimated for the Maunder Minimum (Figure 2.3).

Future

Current ability to predict solar activity is at best primitive. Statistical methods predict sunspot numbers and the 10.7 cm radio flux 12 months in the future with moderate success. There are also precursor methods that predict the strength of the next solar cycle from the behavior of polar structure on the Sun and geomagnetic activity in the declining phase of the current cycle (e.g., Schatten and Pesnell, 1993; Thompson, 1993). But there is limited physical understanding of why these precursor methods should be appropriate except that the magnetic fields and corona near the solar poles change near solar maximum and hence may herald the onset of the new cycle before the next generation of sunspots appears.

On century time scales, the periodicities of 11 and 88 years identified in the sunspot record, together with the 208 year periodicity found in the ^{14}C record, provide limited guidance to future solar behavior, such as the occurrence of the next Maunder Minimum. The time span of solar measurement is simply too short for reliable prediction of solar extrema occurring sporadically every 200 years or so. Nevertheless, it has been speculated that the concatenations of the 208 and 88 year periods may have contributed to generally increasing solar activity levels during the twentieth century, with maximum activity predicted to occur during the first half of the twenty-first century (Damon and Sonnet, 1991).

Predictive capability will be substantially improved when a complete understanding is obtained of the mechanisms within the solar atmosphere that produce the emitted radiation, and form sunspots and plages. Predicting the Sun-as-a-star energy quantities needed for global change studies will ultimately require development of a theory for the solar dynamo that can accommodate known solar behavior. Yet, the very nature of solar variability, whether driven by an internal chronometer (Dicke, 1978) or by stochastic or chaotic processes (Mundt et al., 1991; Morfill et al., 1991; Kremliovsky, 1994), remains elusive. Solar activity levels may well defy reliable prediction in the near future.

7

Research Strategies

The most urgent need for determining solar influences on global change is reliable, continuous monitoring of solar irradiance over many decades. Because of the lack of calibration accuracy of existing solar radiometers, acquiring a record of solar forcing suitable for global change research will require continuous monitoring by multiple spacecraft with sufficient temporal overlap to ensure long term precision by transferring calibration accuracies. Effort is also needed to improve the long term precision and the calibration accuracies of existing instruments.

To fully address the role of solar influences in global change, additional research will be needed to augment the solar monitoring. In particular, the terrestrial effects of solar forcing (from the top of the atmosphere to the surface of the Earth) must be continuously ***monitored***, also over many decades, and an ***understanding*** developed of the physical feedback mechanisms responsible for these effects. Ultimately, the ability to ***predict*** past and future solar influences will derive from improved knowledge of the origins of solar variability.

This slate of activities, in the broadest sense, encompasses much of the domain of solar-terrestrial relations. Indeed, our current knowledge of solar influences on global change has been derived, for the most part, not from research with this specific goal but from core solar and atmospheric research programs that should continue to be supported.

However, core research has tended to focus along classical disciplinary lines rather than on the coordinated, long-term, cross-disciplinary monitoring activities that are essential for documenting how and why the Earth's environment changes. The fundamental sources of information will be data bases that are built up slowly over relatively long periods, and the interpretation of the changes that occur will involve cross-disciplinary analyses that use information from many such data bases. To be effective, global change research must transcend existing disciplinary barriers and encourage interactions that cross disciplinary lines. Such interactions are currently difficult to establish.

This chapter assesses specific ongoing and planned research activities most relevant to solar influences on global change, and then discusses some programmatic issues. The core research programs that exist at present neither accommodate nor foster cross-disciplinary global change research needs. Measuring and modeling the variations in energy input from the Sun to the Earth is essential for research on solar influences on global change. But it is not a prime goal of existing or planned solar physics research, since knowledge of solar processes is better achieved with highly spatially resolved observations of portions of the solar disk. Nor is it a prime goal of Earth science research, for which it is an initiator but not an indicator of the physical processes of interest. As a distinct cross-disciplinary task, the study of solar influences on global change is championed by neither the Earth science nor the solar astrophysics community.

MONITORING SOLAR FORCING

Reliable measurements of solar energy inputs to the Earth system extend over less than 20 years (which is less than two solar activity cycles). Existing measurements indicate significant variability of essentially all solar parameters on essentially all time scales, from minutes to decades. In acquiring a suite of solar irradiance measurements with sufficient long term precision for global change research, important aspects of space based solar metrology obtained from the experiences of the 1980s must be used to guide research strategies for the 1990s and beyond.

Existing solar radiometers have sufficient short term precision to measure irradiance variations generated by solar rotation over time scales of days and weeks, but the precision of the measurements over the 11-year activity cycle is much less secure because of instabilities in radiometric sensitivity. Critical is the recognition that true solar irradiance variations cannot be reliably determined from successive measurements by different instruments unless they can be intercompared via overlapping flight epochs. This is because the systematic errors in current state-of-the-art solar radiometric metrology are of the order of the solar cycle variability itself. Improvements in instrument precision and calibration accuracy are thus important. Furthermore, it must also be recognized that because satellite instruments can (and do) fail, concurrent measurements by at least two instruments are essential to ensure the continuity of the data base. Previous studies by the National Academy of Sciences (1988, 1991) have also emphasized the need for overlapping data bases.

Total Solar Irradiance

The detection of solar luminosity variability during solar cycles 21 and 22, and the interpretation of this variability in terms of solar magnetic activity, thus far underscores the need to extend the solar irradiance data base indefinitely with maximum possible precision. The data are needed for the forseeable future to reduce the uncertainties in the detection of anthropogenic climate forcing. A careful measurement strategy will be required to sustain adequate precision ($<$ 50 ppm, or 0.005 percent). Due to the likelihood of instrument degradation, solar monitoring experiments using current radiometric technology can be expected to last no more than one decade. Drifts in sensitivity throughout a 10-year mission must also be anticipated and detected. Data gaps through instrument failure must also be prevented. Therefore, adequately overlapping experiments and intercomparison of successive experiments is crucial.

Continuation of the total solar irradiance data base that extends from November 1978 to the present is in serious jeopardy. The current and proposed total solar irradiance monitoring program shown in Figure 2.1 relies almost exclusively on one upcoming NASA mission, and as presently conceived will not satisfy the requirement for continuous overlapping experiments, nor even for third party comparisons between successive experiments. ACRIM III has been selected for inclusion on the Earth Observing System (EOS) CHEMISTRY 2003 platform, but this is not scheduled for launch until early in the twenty-first century. Thus it is very unlikely that the requisite overlap between UARS and EOS will be achieved, let alone the multiple measurements crucial for data validation and data loss prevention. Uncertain funding prospects during the next decade have already threatened removal of ACRIM III from EOS (Hartmann et al., 1993). There are no plans for subsequent solar irradiance measurements.

The European Space Agency (ESA) will contribute to the total solar irradiance data base with an experiment on the Solar and Heliospheric Observatory (SOHO) Mission to be launched in mid-1995, which may be able to provide a third party comparison between UARS and EOS total irradiance data, although this will require that the SOHO mission be extended significantly beyond its planned lifetime.

Achieving a meaningful third party comparison with radiometers flown on the Space Shuttle will be difficult, if not impossible. Improved precision may be achieved by deploying the new cryogenic radiometric technology (Foukal et al.,1990); while it requires expendable cryogens at present, and is therefore limited to recoverable or serviceable platforms, it could provide a useful backup. Aside from whether long term instrument precision can be demonstrated from one shuttle launch to the next, despite present absolute uncertainties of more than ± 0.2 percent, is the impact of significant rotational modulation. Solar irradiance can change by as much as 0.3 percent per 13 days, precluding the reliable determination of long term trends of 0.1 percent per decade from sporadic measurements for a week or so, once per year.

Rather than including essential solar monitoring instrumentation as part of complex space platforms that inevitably suffer delays, a series of small, overlapping satellite missions dedicated to monitoring solar irradiance variability will likely prove to be a more reliable strategy for obtaining the requisite data for global change research.

Solar Spectral Irradiance

Obtaining an unbroken, reliable record of the Sun's UV irradiance variations will require an approach similar to that identified for total irradiance monitoring. This includes utilization of the overlap measurement principle whereby new instrumentation has sufficient temporal overlap with the instrumentation that it is intended to replace. Having at least two simultaneous measurements of solar spectral irradiance at any one time will provide this overlap as well as prevent a break in continuous data should one of the instruments fail. The sensitivity of space borne solar instrumentation must also be tracked throughout operation.

The proposed UV irradiance measurement scenario is illustrated in Figure 3.2. The demise of the SME spacecraft precluded overlap with measurements from UARS of solar spectral irradiance from 115 to 410 nm. Beyond the UARS mission (i.e., solar cycle 23 and subsequent cycles), it is presently planned to launch a second SOLSTICE instrument as part of the EOS CHEMISTRY 2003 platform in the early twenty-first century. Continuity between the UARS solar UV spectral irradiance data base and EOS is critical but improbable, either by direct overlap or third party comparisons. Uncertain funding prospects during the next decade also threaten removal of SOLSTICE from EOS. No additional space borne UV spectroradiometers are planned for overlap or backup or for missions beyond EOS.

NOAA's SBUV/2 instruments are currently measuring solar UV irradiances, but only as a subset of their primary goal of measuring atmospheric ozone and only at wavelengths longer than 160 nm. Furthermore, the SBUV/2 instruments lack the capability of end-to-end in-flight sensitivity tracking, making it difficult, if not impossible, to adequately account for instrumental effects in the data. It is unlikely that measurements of solar UV irradiance made from the Space Shuttle will be adequate for assessing solar influences on global change during the coming decades. They are unable to quantify the contribution of short term irradiance variability to measured long term trends. With measurement uncertainties of about 5 percent in the region between 200 and 300 nm, which is thought to be most important for global change, they are insufficiently accurate, since this is the order of the solar cycle UV irradiance variability.

Because of the paucity of plans for future solar UV spectral irradiance measurements, the preferred strategy would be to include instruments to measure both spectral and total irradiance on a series of overlapping solar monitoring satellites. If possible, instruments that measure changes in the solar spectrum at wavelengths longer than 400 nm, for example in the region between 1 and 4 microns, should also be considered. Despite the importance of this latter infrared spectral region for the biosphere, the solar cycle changes in this spectral region have yet to be determined and may even be out of phase with the sunspot cycle (see Figure 1.1).

An opportunity exists to obtain new measurements of solar EUV and X-ray spectral irradiances with the Solar EUV Experiment (SEE) instrument that has been selected for inclusion on TIMED (Woods et al., 1994) as part of NASA's new Solar Connections Program. Limited solar EUV monitoring will also be provided by SOHO. An example of a spacecraft program for developing the necessary understanding of the EUV irradiance variations is SOURCE (Solar Ultraviolet Radiation and Correlative Emissions). The intent of this mission would be to measure EUV spectral irradiance in conjunction with observations by full disk solar imagers to record EUV emissions from solar surface magnetic structures, with the goal of developing reliable empirical (but physically based) irradiance variability models for use in studying solar forcing of the upper atmosphere (Smith et al., 1993).

Even if adequate flight opportunities can be established for solar irradiance instrumentation, a data base with sufficient accuracy for establishing the variability of solar UV energy inputs to the Earth will only be achieved through a continued commitment to innovative radiometric programs dedicated to the improvement of absolute measurement accuracies. Needed are extensive absolute radiometric cross-calibrations using a variety of laboratory irradiance sources, verification and improvement of the irradiance and detector standards maintained by the National Institute of Standards and Technology (NIST), laboratory intercomparisons of the flight instruments, and provision for end-to-end calibration monitoring during integration and flight.

Energetic Particles

The observational needs for energetic particles are long term monitoring of relativistic electrons and particle fluxes, with more emphasis on the higher energies ($>$ 100 MeV) than has been the case in the past. The recently launched UARS and Solar, Anomolous and Magnetospheric Particle Explorer (SAMPEX) missions are currently measuring some particle energy inputs, and further measurements are planned for the upcoming NASA International Solar-Terrestrial Program (ISTP)/Global Geospace Study (GGS) and the continuing Department of Defense (DoD) Defense Meteorological Satellite Program (DMSP). However, simultaneous measurements of the atmospheric response to these inputs will only be made from a few selected ground stations as part of the National Science Foundation's (NSF) Coupled Energetics and Dynamics of Atmospheric Regions (CEDAR) program. Additional measurements of energetic particle input and the atmospheric response should continue to be made with the suborbital and NOAA programs.

Ground Based Solar Variability Indicators

A number of existing ground based solar observing programs have proven, over the past decade, to be extremely valuable and cost effective for studying solar influences on global change, and these programs should be continued indefinitely. Two types of ground based solar observations are important: 1) measurements of the relative solar flux in specific regions of the visible, infrared, and radio spectrum that reflect the integrated variations caused by magnetic brightness sources in the solar atmosphere, and 2) spatially resolved observations of the solar disk from which the active region features that contribute to the irradiance variations can be identified and characterized. These data (Figure 6.3) are utilized to generate surrogates for solar activity modulation of the total solar irradiance (Figure 6.4) as well as for the UV and EUV fluxes. The 10.7 cm radio flux, available since 1947, is a uniquely long solar flux time series which must be continued.

Of the ground based solar observing programs, perhaps the most important is the NSF-funded program at the National Solar Observatory (NSO) Kitt Peak and Sacramento Peak facilities for measuring spectral irradiances of chromospheric lines (He I 1083 nm and Ca II K) and coronal lines (Fe IV 530.3 nm) in units relative to the background continuum. The NSO data base is especially valuable because it now extends for 20 years, from 1974 to the present, longer than any of the continuous satellite records of irradiance variations. Extending the data base will further increase the possibility of statistically meaningful correlations with climate parameters. The usefulness of the NSO ground based solar monitoring for global change research could be enhanced by making measurements on a daily basis and with photometric calibration. Improvements such as these have been detailed in the Proceedings of a Workshop on Solar Radiative Output Variations (Foukal, 1987) and incorporated in NSF's Radiative Inputs of the Sun to the Earth (RISE) initiative.

A number of solar observatories record full disk solar images detailing magnetic field strengths and photospheric and chromospheric magnetic structures. These measurement programs provide basic data (frequently daily) on the evolution of the details on the solar surface. With improved analysis, these images could provide quantitative characterizations of the active regions that generate solar energy output variations and that are needed to construct full disk irradiance variations.

Also relevant are observations of variability in Sun-like stars. Existing programs now include data collected over more than a decade; they should be continued indefinitely. The use of this larger stellar sample can yield important clues about solar variability on otherwise inaccessible time scales, such as, for example, the propensity of brightnesses of stars similar to the Sun to oscillate regularly or irregularly. Where possible, spectral irradiance observations should also be made -- for example, observations of the UV emission of these Sun-like stars.

Indirect records related to solar variability have been provided for many decades by measurements of the Earth's magnetic field from global ground based magnetometer networks. In many cases, these ground based data have been used to construct global geomagnetic indices, represented, for example, by Kp, Ap, and Dst indices, that describe amplitude and directional changes in the varying field of the Earth. As noted by NAS (1988), because ground based geomagnetic activity indices extend continuously from before the space age to the present, they are of great importance to long term solar-terrestrial research. Ground magnetometers also provide the long term record used to monitor the secular variation of the Earth's internally generated magnetic field. Maintaining the ground based magnetic records entails keeping in operation a sufficiently dense and properly distributed network of ground magnetic observatories.

MONITORING TERRESTRIAL SOLAR EFFECTS

Lower Atmosphere

Continuous monitoring of the solar irradiance over the next several decades, as discussed in the previous section, would provide the opportunity to quantify its potential impact on the climate system, assuming that observations of climate and other potential forcing mechanisms (trace gases, aerosols, ozone, etc.) are maintained as well. In the lower atmosphere, it is expected that an increase in solar radiation will, like increasing greenhouse gases, warm the Earth's surface. In the stratosphere, however, the two effects would produce temperature changes of opposite sign. Augmenting long term observations of tropospheric parameters with similar observations of stratospheric parameters would constitute a monitoring program that could separate these diverse climate perturbations and help isolate a greenhouse footprint for climate change. Monitoring global change in the troposphere and understanding climate forcings and feedbacks is the specific focus of the Climate and Hydrologic Systems science element of the USGCRP and a key element of its other facets, not just of the study of solar influences on global change. The need to monitor the stratosphere is also important for global change research in its own right within the Biogeochemical Dynamics science element, because of the existence of the stratospheric ozone layer, and this is discussed more extensively in the following section.

Global monitoring of the solar ultraviolet radiation reaching the biosphere is relevant for studying solar influences on global change as well as for the Ecological Systems and Dynamics science element of the USGCRP; both involve changes in middle atmosphere ozone. A much needed step toward a credible ground based UVB measurement program is the development of absolutely calibrated UVB photometers and high resolution spectrometers, presently being sponsored by the Department of Agriculture. While improved monitoring techniques are being developed in the U.S. by programs within the Department of Agriculture, the Environmental Protection Agency (EPA), and NIST and by similar endeavors in other countries, these capabilities will still not meet all of the needs for global monitoring. Neither the accuracy nor the adequacy of the measurements will be sufficient to determine global anthropogenic trends or natural solar-induced cycles, especially over oceans. At selected sites, accurate, detailed measurements of the UV radiation as a function of wavelength are needed to verify the monitoring program and for model verification studies. Also necessary at these sites are accompanying measurements of other parameters, such as transparency, total ozone, and aerosol loading. If techniques for determining incident UV radiation from space can be developed, these satellite data in combination with the ground based measurements could provide the needed monitoring capability.

Middle Atmosphere

Both in situ and remote measurements of as many of the chemical constituents in the middle atmosphere as possible are needed to understand the processes affecting ozone and other important constituents. Achieving these measurements is a focus of the Biogeochemical Dynamics science element of the USGCRP, and a substantial effort in this area is underway, much of it embodied in the National Ozone Plan. Such missions as the Upper Atmosphere Research Satellite (UARS) and the implementation of the ground-based Network for Detection of Stratospheric Change (NDSC) should add greatly to our knowledge as well as provide a suite of middle atmospheric observations that should better define the response of the middle atmosphere to solar forcing.

However, it must also be recognized that determining solar influence on the middle atmosphere requires global measurements over not just one but several solar cycles, at least. In addition to ozone, measurements of nitrogen oxides (NO, NO_2, HNO_3) are particularly important in determining the influence of solar radiation, energetic particle, and cosmic ray variations over the 11-year solar cycle. NOAA continues to monitor atmospheric ozone through its operational polar orbiting satellite series (albeit with uncertainties arising from the instrument diffuser degradation), and calibrated measurements of individual profiles are being made by the Stratospheric Aerosol and Gas Experiment (SAGE) II.

It is doubtful that continuous measurements of the parameters necessary for adequately assessing solar influences on the middle atmosphere will continue beyond the UARS time frame. This region of the atmosphere is not the focus of EOS. With the continuing reduction in scope of the EOS program, there is no assurance of continuous records of certain species measured by UARS -- such as temperature, O_x, NO_x, HO_x, and Cl_x -- that are key to the long term data base needed for global change studies. Some critical constituents, such as OH, are not even measured by UARS; the only measurements of OH planned will be from a few Space Shuttle flights of the Middle Atmosphere High Resolution Spectrographic Instrument (MAHRSI) experiment. Ground based networks, such as NDSC are therefore of major importance.

A program that combines both satellite measurements for global coverage and supportive suborbital data is needed to guarantee the science, calibration, collaboration, validation, and in situ studies of specific middle atmospheric processes relating to energetic particles. Neither is in place. The UARS Particle Environment Monitor (PEM) experiment and the SAMPEX mission are providing some information on particle energy deposition to the middle atmosphere that may allow the relevant particle energy inputs to be sufficiently well defined that proxy measurements by the Geostationary Operational Environmental Satellite (GOES), DoD and ISTP programs will be adequate for providing the long term data base.

Upper Atmosphere

Like the state of knowledge of the relevant solar energy inputs, global information on the responses of the thermosphere and ionosphere system to solar forcing faces a dearth of observational programs. Different types of measurements are needed to elucidate aspects of the upper atmosphere relevant to global change: its extensive, natural, solar-driven variability, the predicted anthropogenic forcing, and the coupling of changes of both natural and human origin to the lower layers of the atmosphere. The emphasis of these measurements must be on a global spatial scale and for long time scales, with the goal of defining the present structure from which future trends in global change can be derived.

No existing or planned long-term spacecraft programs address any of these issues. NASA is currently pursuing initiatives to launch an extensive investigation of the upper atmosphere with the Thermosphere Ionosphere Mesosphere Energetics and Dynamics (TIMED) Mission as part of its Solar Connections Program. While the chief goals of the program are oriented toward process studies rather than global change issues, the observations should provide important new knowledge about the mesosphere and upper atmosphere. Observations are planned by the DoD Remote Atmospheric and Ionospheric Detection System (RAIDS) spacecraft experiment during the present decade, if the Space Test Program can identify a suitable launch vehicle. But neither RAIDS nor the limited upper atmosphere data from UARS, by themselves, will be sufficient to detect and understand processes important for global change.

A long term data base may become available from UV remote sensing instruments planned for DMSP satellites beginning in the late 1990's. While the goal of these sensors is to provide real time monitoring information on thermospheric and (F-region) ionospheric weather, the data base should extend beyong a decade. However, with the planned merge of DMSP and NOAA weather satellites, a final payload configuration has not been defined.

The dearth of long term studies of the Earth's upper atmosphere from space platforms places greater responsibility on ground based observations and in particular on NSF's CEDAR program, which was originally identified with NSF's Global Geoscience Program and is, again, principally focused on process studies. Thus the activities pursued as part of CEDAR should be extended enough to study the long term and global scale phenomena important for understanding and predicting global change. In particular, cooperative measurements by the CEDAR community could enhance the scientific products of the UARS, RAIDS, and TIMED missions.

UNDERSTANDING SOLAR INFLUENCES ON GLOBAL CHANGE

Current understanding of solar influences on global change is largely the result of studies initiated by a few interested researchers, working largely within their own institutions, on problems ancillary to those of global change, and without a broad community base. The disciplinary differentiation between middle and upper atmosphere and climate studies limits the opportunities for combined-region research in this area (and others). In addition, the traditional skepticism that often greets solar/weather relationships means that such studies generally receive low priority. Existing atmospheric and climate modeling programs should be encouraged to become more integrated and to give higher priority to solar forcing in addition to anthropogenic effects that are their primary focus.

Studies of Present Day Behavior

Proper specification of the response of the climate system to contemporary solar forcing requires the identification of all relevant pathways and feedbacks operable over decadal time scales. This will ultimately require combining climate and middle atmosphere models to construct extended three-dimensional models with complete, coupled radiation, chemistry and dynamics. Investigations using improved models may then lead to new understanding of the potential solar cycle/weather relationships that are not simulated by existing climate or middle atmosphere models alone.

Given that the most direct impact of varying solar radiation is on the middle atmosphere, there must continue to be detailed observational and modeling studies of the relationship of the solar cycle to middle atmosphere phenomena, such as stratospheric warmings and mean wind and temperature fields. The effects must then be followed into the troposphere -- for example, alteration of planetary wave energy propagation from the troposphere by modification of zonal winds in the stratosphere. Direct and indirect tropospheric effects, such as the relationship of solar activity to cirrus clouds, should also be pursued. Statistical tests of the relationships must be continued, and the perspective of the NOAA Climate Analysis Center, with its solar cycle-influenced forecasts, should be brought to bear.

The relationship of the QBO to the whole question has to be studied, both through the observation of altered energy pathways in the troposphere associated with the changed tropical lower stratosphere zonal winds and through its interaction with solar cycle effects. GCMs of the troposphere and stratosphere should be used in combination to model the direct solar cycle influence, the direct QBO influence, and the combined processes. The ability of models to reproduce results such as those shown in Figure 2.6 would not only validate the solar cycle relationships, but would go a long way toward improving our confidence in the capability of models to predict the regional effects of global change.

Atmospheric modeling studies of trends in ozone, temperature, and winds need to more accurately consider the combined effects of variations in solar flux, cosmic ray, and energetic particle influences, their transport to regions far from where the energy is deposited, and the feedback mechanisms that they invoke. The processes by which changing solar energy inputs impact the lower stratosphere, as implied by observations, require particular attention, as do the influences of solar soft X-rays and energetic particle precipitation on the production of nitrogen oxide molecules. These studies are essential to establish a solid definition of the background against which human-related effects on the atmosphere need to be measured. As noted previously, observations of middle atmosphere trends may allow separation of solar from anthropogenic forcing in the troposphere.

With respect to understanding solar influences on the upper atmosphere and possible indirect forcing of global change, numerical models of physical and chemical processes and global circulation of the coupled thermosphere-ionosphere-mesosphere system should be developed and used to investigate the effects of solar terrestrial couplings on global change. These models should also consider the effects of anthropogenic forcing, such as CO_2 and CH_4 increases, on the properties and dynamics of the upper atmosphere. Techniques for coupling these models to models of lower atmosphere chemistry and dynamics should also be explored.

A quantitative global climate model of the entire coupled Earth system is likely to be needed to understand the global changes expected in the twenty-first century, despite the difficulty of constructing such a model. Examples of the types of studies that could be pursued are investigation of the possibility that information on the chemical coupling between the upper atmosphere and the biosphere on historical time scales might be revealed by deposits of odd-nitrogen species in ice cores (a topic that has obvious links to the Earth System History science element of the USGCRP), and the extent of dynamical coupling that might generate solar/QBO signatures common to a wide range of altitudes within the atmosphere.

A number of laboratory studies are needed to support global change research. Photodissociation processes play an important role in determining the production and loss rates of many of the constituents in the middle and upper atmospheres. However, there are few observational data to verify the photodissociation rates determined and used in models of atmospheric photochemistry. Relevant laboratory studies include, for example, measurements of the O-CO_2 vibrational exchange coefficient, of the branching ratio of N_2 dissociation by electron impact to $N(^4S)$ and

$N(^2D)$, and of the rate coefficients of a number of reactions involving metastable species.

Records of the Past

It is clearly necessary to understand natural variability and other forcings, including possible solar influence on the decadal/century time scale, to be able to understand and predict the likelihood of anthropogenic-induced climate change in the next several decades. Concerning the influence of solar variability and past climate, much needs to be done from both the observational and the modeling perspectives. Clearly, there is significant overlap in the focus on paleo- and recent climate with similar goals of the Earth System History science element of the USGCRP.

For relatively recent climate variations, such as the purported Little Ice Age, more quantitative and global coverage of the climate (temperature in particular) is obviously needed. This will require a program to use and combine many paleoclimate indicators, such as those derived from tree rings, glaciers, pollen, and corals. Surrogates for solar irradiance variations are needed here in particular, if any connection between the decadal to century scale climatic oscillations and solar variability is to be proven. Models of the climate system should be used with potential solar irradiance (and other) perturbations, and compared with paleoclimate observations over different centuries. Both the magnitude and the spatial patterns of the effects will help in assessing the likely solar irradiance contribution and model sensitivity.

A joint NSF/NOAA funded program on the Analysis of Rapid and Recent Climatic Change (ARRCC) is in progress, with the intention of using all available climate indicators to develop global climate assessments for several cold epochs in the eighteenth and nineteenth centuries. Combined with estimates of potential forcing factors (e.g., solar variability, volcanic aerosols, and ocean circulation changes), modeling studies will attempt to reproduce the observations, perhaps implicating one or another of the mechanisms. Modeling studies are already under way to assess the impact of recent solar variations and should be carried to fruition.

From the paleoclimate perspective, orbital-induced insolation variations are now thought to serve as the principal pacemakers of glacial cycles. Nevertheless, the mechanisms through which changes in the distribution of insolation can influence climate are as yet undefined and controversial. The long term climate change proxies from marine data were found quite by accident to exhibit orbital periodicities, contrary to the prevailing view of most climatologists. Periodicities and phase lags in the climate system during the Pleistocene need better definition. For interpreting the orbital-induced insolation variations, it is important to clarify the absolute dating of the paleoclimate record, including the deep sea record, without the aid of assumptions on orbital parameter influence. Additional calcite veins from as diverse a geographical area as possible are needed to better understand the local/global signals as well as hydrologic cycle influences. It is also necessary to assess the newly available ice core data, including the critical comparison of Greenland and Antarctic results.

From the modeling end, additional GCM experiments should be made to quantify the solar insolation changes needed to support low elevation ice sheets. The models used must have sufficient horizontal and vertical resolution to address the problem of surface layer effects in specific areas and must be able to produce a reasonable cryospheric climatology for the current climate (i.e., correct seasonal variation of snow cover and reasonable mass balances for current ice sheets). These modeling experiments should be encouraged to explore all possible feedback mechanisms whereby solar radiation changes may influence global climate. As orbital variations represent our most quantifiable solar insolation changes, they provide tools both for quantifying climate sensitivity and for validating climate models on long time scales.

UNDERSTANDING AND PREDICTING SOLAR VARIABILITY

Acquiring a reliable data base of solar energy inputs to the Earth through the next decades is essential for monitoring and understanding solar forcing of global change. Understanding and predicting solar forcing on the much longer time scales needed for global change research will require additional effort. In particular, the origins of solar energy variations must be understood in terms of the variable Sun to allow historical solar variability to be reconstructed from proxy records, for tying contemporary solar variability measurements to each other and to stellar analogs, and for forecasting irradiance variations expected to arise from solar magnetic activity.

It is hoped that the acquisition of a reliable, long term data base of solar irradiance measurements from the UARS and Yohkoh spacecraft will facilitate improved understanding of the origins of the irradiance variations through analysis of these data in conjunction with auxiliary solar data, such as spatially-resolved magnetic field maps, and Ca II and He I images and fluxes. The irradiance variations must be physically connected to the fundamental cause of solar variability, which is solar magnetic activity, to achieve the ultimate goal of prediction.

Until the data base of solar EUV spectral irradiance observations has been augmented substantially, significant improvements in the simple empirical models that predict EUV irradiance variations from solar activity proxies will not be possible. These models are derived from correlation analysis of a chosen solar proxy with existing data and are thus constrained by the inadequacies of those data. Some progress may be possible by developing more sophisticated models based on the physical solar processes that drive solar irradiance variability. By incorporating independent knowledge of the characteristics of active region emission and other magnetic features on the solar disk, such models can potentially provide improved predictions of irradiance variability as well as a tool for investigating the origin of the variability. Concepts for models of this latter type have recently begun to emerge, but are currently unfunded, partly because studying the variability of the Sun-as-a-star has not enjoyed high priority focus within the solar physics research community.

To reconstruct past changes in solar radiative energy inputs to the Earth, it is necessary to determine the validity of empirical models relating irradiance variations to surrogate phenomena of solar activity. Existing empirical models (see Figure 6.4), although essential for verifying and interpreting the variations recorded by satellite instrumentation, are nevertheless rudimentary. Still needed, for example, is the ability to quantify the solar irradiance variations of the past several thousand years; this will necessitate the incorporation of solar activity indicators, such as tree-ring ^{14}C or ice-core ^{10}Be, and studies to define their physical relationships to changes in the Sun's radiative input to the Earth.

For understanding the origins and mechanisms of solar radiative output variations, an important proposed research program is RISE (Radiative Inputs of the Sun to Earth) which includes both space based and ground based measurements and analyses that must be the foci of solar variability studies from a global change perspective. With the exception of some NSO ground based observation programs, RISE is the first research program with the goal of understanding the variations of the Sun-as-a-star. One component of RISE is now a program in the global change initiative at NSF within the USGCRP. This includes an effort to obtain precision photometric images of the solar photosphere and chromosphere using a dedicated basic telescope system designed specifically for photometric observations. If adequately funded, RISE could also provide future support for the analysis and interpretation of the historical solar image data base of contemporary ground based data relevant to understanding the Sun as a variable star and of the terrestrial pathways and processes through which solar variations might impact global change.

The ongoing Global Oscillation Network Group (GONG) helioseismology program (Harvey et al., 1987) continues to be a high-priority international effort that is also related to the study of solar variability. The National Solar Observatory directs this effort to build, install, and operate optical helioseismographs at a network of sites around the world. This project will yield almost continuous observations of velocity and brightness oscillations over the full solar disk starting in 1994, and promises a major advance in understanding the structure and dynamics of the solar interior, where solar activity is thought to be generated and modulated.

Experiments on the SOHO spacecraft will also provide valuable helioseismic and solar atmosphere structure measurements beginning in 1995. If they can be made with sufficient long term precision, measurements of the solar diameter may also prove beneficial, insofar as they serve as a proxy for other significant solar changes. Further research is needed not only to measure and model solar processes on the fundamental scale of the magnetic flux tube, but also to convert the actual magnetic fluxes into radiative and particle outputs from the full solar disk. The Mechanisms of Solar Variability (MSV) program, a recently conceived Solar Research Base Enhancement within NASA's Space Physics Division, may provide some progress in this area, providing that the proposed high spatial resolution investigations are demonstrated to be directly relevant to understanding disk-integrated solar emission variations.

PROGRAMMATIC APPROACH

Need for Interdisciplinary Efforts

Clearly no physical walls separate the various parts of the coupled Sun-Earth system -- the heliosphere, the magnetosphere, the ionosphere and upper atmosphere, the middle atmosphere, and the climate system -- from one another. However, separate disciplines have developed over the past several decades to study the various parts of this system. This situation has led to intellectual and administrative walls delineating distinct scientific communities that make it difficult to study the entire coupled system.

Yet to successfully investigate the influence of solar effects on global change requires a program that compasses all of these areas of research. The need for this interaction is clearly demonstrated by the following hypothetical examples. A search for connections between solar and atmospheric (or oceanic) behavior might first proceed with correlation studies between some broad indicative measures -- for instance the solar radio flux at 10.7 cm and terrestrial surface temperature. A next step, though, might be to look for correlations between parameters hypothesized to be involved in mechanisms for such effects -- for instance, cosmic rays and cloudiness. A parallel effort would comprise cloud physics experimentation to see if some of the hypothesized crucial cloud nucleation mechanisms actually take place in the laboratory. These steps involve at least four separate sub disciplines.

As another example, a search for the effects of solar UV variations on the atmosphere requires a model properly formulated to include both direct and indirect UV effects on the middle atmosphere the lower atmosphere and the couplings between. This in turn calls for interdisciplinary collaboration.

Clearly, then, for research in this area to succeed, scientists in various disciplines need to focus some of their activities on this specific area of research and to interact in formulating research approaches.

Connections to Other Areas of the USGCRP

Because the Sun is the dominant source of energy for the Earth, the need to understand solar influences on global change pervades almost all other areas of the USGCRP. This is illustrated in Table 7.1. It is clear that solar influences have the potential to affect Climate and Hydrological Systems. As has been mentioned, U.S. seasonal winter forecasts are already being implemented with consideration of solar influences. There is also a clear relationship to Biogeochemical Dynamics. Solar variations are known to affect the middle atmosphere, and these effects must be considered when looking for trends in stratospheric ozone. There is an obvious relationship between solar influences on global change and Ecological Systems and Dynamics. Solar effects on climate, if shown to be important, can have ecological impacts, and of course solar effects on stratospheric ozone play a role in modulating the UV-B flux into the biosphere. In Earth System History, the record of past solar variations must be considered. Less direct, but possibly present, are solar influences on Human Interactions through the climate connection, the impacts of changing UV radiation on human health, the disruption of society caused by power and communication failures and the reliance of society on satellite technology to meet many needs, including defense.

Agency Roles

The USGCRP is an attempt to better coordinate research on global change among various U.S. agencies. For instance, NOAA, NASA, NSF, and DoD all have research activities involving the Sun and aspects of its influence on the Earth system. In fact, the products of the basic research programs funded by the various agencies have provided the basis for current understanding of solar influences on global change. These activities have quite different goals, however, even within different programs in an individual agency. Improved coordination is needed within and among U.S. agencies to focus existing research and to more effectively direct new research on solar influences on global change.

NOAA, for example, has traditionally been concerned with the long term monitoring aspects of the problem. A substantial data base of energetic particle fluxes has been built up from particle monitors on its

Table 7.1 Connections between Solar Influences on Global Change and other science elements identified by the U.S. Global Change Research Program.

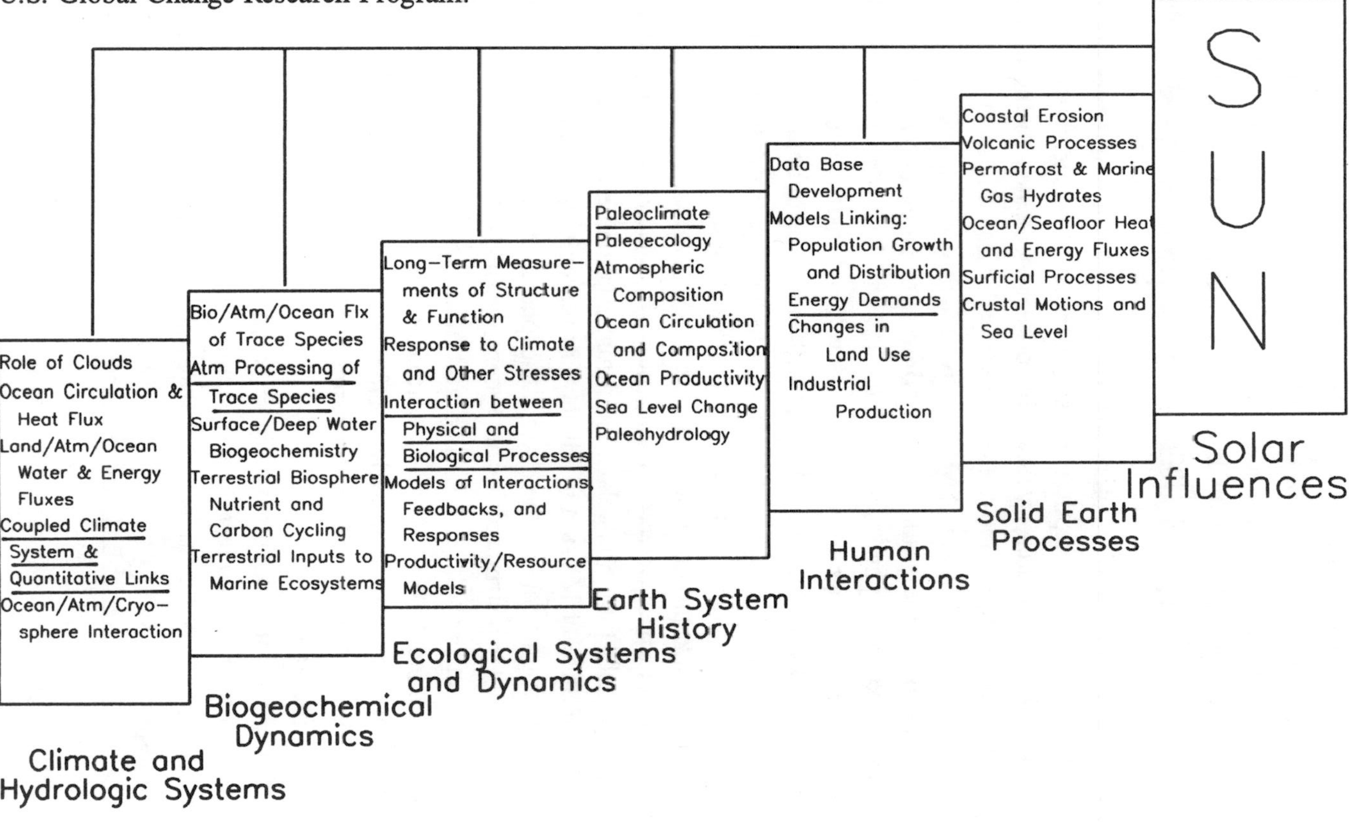

operational satellites, and routine monitoring of solar irradiance has been carried out by the ERBE and SBUV/2 instruments, albeit with less than adequate temporal resolution (in the case of ERBE) and long term instrument monitoring (in the case of SBUV/2). Since NOAA's operational satellite program is expected to continue indefinitely, it could provide the logical platform for many of the monitoring activities discussed above. NOAA's global change research program has been substantially enhanced since the inception of the Climate and Global Change program, and enhancement of the solar component of that program would be appropriate.

NASA, since its inception, has supported solar and terrestrial research. Several NASA research satellites have been dedicated to these areas of research; however, the most recent mission dedicated to solar research, the Solar Maximum Mission (SMM), was launched a decade ago and ceased operation fully five years ago. In contrast to NOAA, NASA's programs in solar and space physics are oriented toward short term research; that is to say, NASA's programs use newly developed instrumentation that gives new types of data for a time but with little long term commitment. Even within NASA, different research divisions have quite different approaches to the study of solar influences on the Earth system. The main part of solar and space physics resides within NASA's Space Physics Division in the Office of Space Science. Here, the main concern is to understand the workings of the Sun-solar wind-magnetosphere-ionosphere system. NASA's Office of Mission to Planet Earth monitors the Sun with a view to elucidating its role on the lower atmosphere, especially climate change; the three solar instruments on its recently-launched UARS are the prime source of current solar monitoring, and future solar monitoring is planned as part of the EOS. Then, of course, there is the NASA operational interest in predicting solar activity effects (e.g., orbit decay) on spacecraft operations as recognized by the recently formulated Space Environment Effects Program.

NSF has long supported basic research into Solar-Terrestrial Physics -- how solar changes affect the terrestrial environment. NSF also supports theoretical and observational solar research, primarily using ground based techniques. The primary U.S. program for investigating the relationship between solar variability and climate change is the Solar Terrestrial Research Program in NSF's Division of Atmospheric Sciences. Research activities involve solar physics as well as potential solar-induced climate variations on all time scales. Included are paleoclimate investigations involving records in tree rings, and ice cores, as well as Sun-weather relationships. Funding for the core program in this area is diminishing, however, and will need greater emphasis to strengthen the terrestrial component of the program. NSF lists three programs as part of its global change initiative; Coupling, Energetics and Dynamics of Atmospheric Regions (CEDAR); Geospace Environmental Modeling (GEM); and Radiative Inputs of the Sun to Earth (SunRISE).

DoD also has extensive research activities in solar and space physics. These activities involve both theory and ground and space based observations (both from short term research and long term monitoring points of view). The relevance of solar variability for DoD research is that its effects on the Earth environment must be properly taken into account in military operations and particularly in communications and surveillance. A requirement for operational models of upper atmosphere variability to track and predict satellite trajectories, and of ionospheric variability that affects communications, has led DoD to support, over the past decade, what exists today of upper atmosphere research, focusing especially on the solar influences that determine both its neutral and ion composition. The recently implemented Strategic Environmental Research and Development Program (SERDP) encourages use of DoD resources for global change research. In this regard, the Defense Meteorological Satellite Program could provide regular access to space for solar monitoring and global change endeavors.

Because solar influence on global change is a distinctly cross-disciplinary endeavor, cooperative research by DoD, NASA, NSF, and NOAA is essential for mutual benefit to both the USGCRP and the individual agencies. The research effort is hampered by the lack of a lead agency in this area. One challenge to the U.S. program of research into solar influences on global change is to develop a strategy for agency leadership and coordination in pursuing the overall national effort.

International Apects

Two programs within the International Council of Scientific Unions (ICSU) relate to solar influences on global change. One is the Solar-Terrestrial Energy Program (STEP) of the Scientific Committee on

Solar-Terrestrial Physics (SCOSTEP). The STEP program seeks to better understand the coupling of energy and mass throughout the various parts of the solar-terrestrial system. A related ICSU program is Stratospheric Processes and their Relation to Climate (SPARC), an adopted program of the World Climate Research Program (WCRP). Its emphasis is on understanding the role of the stratosphere in the climate system. Both of these international programs include components involving solar influences on global change.

The International Solar Terrestrial Physics Program (ISTP) is a cooperative effort involving the U.S., Japan, and Europe. The program consists of several spacecraft to be launched in the 1990s by NASA, ISAS, and ESA. The overall scientific objectives of ISTP are to develop a comprehensive, global understanding of the generation and flow of energy from the Sun through the interplanetary medium and into the Earth's space environment, and to define the cause and effect relationships between the physical processes that link different regions of this dynamic environment. The ISTP will provide major contributions to the understanding of the energy flow between the Sun and the Earth's magnetosphere, but its principal objectives do not, at present, include study of energy flow into the lower atmosphere.

8

Recommendations

SCIENTIFIC RATIONALE FOR ASSIGNING PRIORITIES

A program of investigation of solar energy inputs into all parts of the Earth system is very much in concert with the goal and objectives of the U.S. Global Change Program. The USGCRP is motivated by the realization that global change can have tremendous impact on conditions essential to life on earth. This realization provides the basis for prioritization among the various components that can be expected to comprise a USGCRP scientific element on Solar Influences on Global Change. Of highest priority are those activities that will be most important for national and international policymaking.

RECOMMENDATIONS

Primary Recommendation

One activity ranks above all others for determining solar influences on global change:

1. Monitor the total and spectral solar irradiance from an uninterrupted, overlapping series of spacecraft radiometers employing in-flight sensitivity tracking.

There is an urgent need to rapidly implement the necessary long term commitment for this monitoring because of the danger that the present monitoring sequence will be interrupted and the long term record invalidated as a result of lack of instrumental cross-calibration. This primary recommendation is particularly challenging and probably will not be achieved because of the dearth of ready access to space.

A series of small spacecraft dedicated to solar monitoring could provide the necessary data. Overlapping observations are required to cross-calibrate measurements by different instruments whose inaccuracies typically exceed the true solar variability. Simultaneous observations from different instruments provide important validation that real variability, rather than instrumental degradation, is being measured and provide the redundancy needed to preserve the long term data base in the case of instrument failure. Improved radiometric long term precision and calibration accuracies would contribute to a more reliable solar forcing record.

In lieu of a spacecraft series dedicated to solar monitoring, it may be possible to use the NOAA or DMSP operational satellites, for which overlapping is a feature of their design.

Additional Recommendations

To augment the prime monitoring task, a suite of efforts from diverse geophysical research fields is needed to achieve the USGCRP objectives of monitoring, understanding, and predicting solar influences on global change. Pursuit of recommendations 2 to 6 is essential to the cross-disciplinary effort needed to reduce uncertainties in knowledge of solar forcing of global change in order to provide a sound scientific basis for policy-making on global change issues. The actions of recommendations 7 to 12 are essential to ensure that complete understanding is achieved of all potential coupling mechanisms.

2. Conduct exploratory modeling and observational studies to understand climate sensitivity to solar forcing.

Implied connections between the Sun and the paleoclimate record (Milankovitch orbital-induced variations and the Little Ice Age) should be fully investigated for the insights they might provide about the sensitivity of the climate system to solar forcing compared with increased greenhouse gases. New knowledge should be incorporated into existing GCMs utilized for climate prediction.

3. Understand and characterize, through analysis of solar images and surrogates, the sources of solar spectral (and hence total) irradiance variability.

The overall goal of this activity is to improve the ability of solar variability models to calculate solar radiative output variations and to provide reliable proxies to bolster the spaceborne monitoring effort. Toward this end, continue, without interruption, to monitor from ground based observatories the relevant proxy data, in particular certain relative spectral irradiances (such as the He I and Ca II indices, and the 10.7 cm flux) and solar images that display magnetic active regions (using, for example, full disk magnetograms, and He I, Ca II, and white light spectroheliograms). Use the improved solar variability models to extend the variability record into the past and to predict limits on future variability. Also important in this regard is connecting the variability sources to the physical solar processes that modulate the ^{14}C and ^{10}Be records.

4. Monitor, without interruption, the cycles exhibited by Sun-like stars, and understand the implications of these observations for long term solar variability.

Tying the calculations of solar radiative output variations derived from solar observations (Recommendation 3) to the broader stellar context will help in this regard.

5. Monitor globally, over many solar cycles the middle atmosphere's structure, dynamics, and composition, especially ozone and temperature.

Long term records of ozone, temperature, and nitrogen oxides are especially important as they may allow the separation of solar from anthropogenic forcing in the troposphere. Solar effects on this region will only be determined from the results of such monitoring.

6. Understand the radiative, chemical, and dynamical pathways that couple the middle atmosphere to the biosphere, as well as the middle atmosphere processes that affect these pathways.

Both modeling and observational studies are needed.

7. Monitor continuously, with improved accuracy and long term precision, the ultraviolet radiation reaching the Earth's surface.

This effort is critical, not only for determining the dosage of UV radiation at Earth, but also because of the dependence of the UV dosage on ozone concentrations, which are affected by both anthropogenic and solar forcings.

8. Understand convection, turbulence, oscillations, and magnetic field evolution in the solar atmosphere so as to develop techniques for assessing solar activity levels in the past and to predict them in the future.

A reliable theory of the solar activity cycle, of longer term variability, and of stellar dynamos in general will require physical descriptions of the processes that successfully reproduce solar phenomena observed over a number of solar cycles. Reliable monitoring of solar diameter could help to understand solar variability processes. The goal is to understand why the Sun varies at all.

9. Monitor continuously the energetic particle inputs to the Earth's atmosphere.

Space based measurements should emphasize the higher energies (> 100 MeV) and relativistic electrons. Understand, through in situ measurements, the relationship of space based measurements to the energy spectrum and fluxes of both solar and galactic energetic particles reaching different altitudes in the Earth's atmosphere.

10. Monitor the solar extreme ultraviolet spectral irradiance (at wavelengths less than 120 nm) for sufficiently long periods to fully assess the long term variations.

These measurements could be accommodated on the dedicated solar monitoring spacecraft identified in Recommendation 1.

11. Monitor globally over long periods the basic structure of the lower thermosphere and upper mesosphere so as to properly define the present structure and its response to solar forcing.

12. Conduct observational and modeling studies to understand the chemical, dynamical, radiative and electrical coupling of the upper atmosphere to the middle and lower atmospheres.

Analysis of solar soft X-ray forcing of nitric oxide levels, with possible inferences for nitrate deposits in ice cores, is an example of such a study. Ultimately, a global model of the Earth system is needed.

References

Allen, J., H. Sauer, L. Frank, and P. Reiff. 1989. Effects of the March 1989 Solar Activity. Eos, Transactions, American Geophysical Union 70:1479-1485.

Allen, M.R., and L.A. Smith. 1994. Investigating the origins and significance of low-frequency modes of climate variability. Geophys. Res. Lett. 21:883-886.

Auclair, A.N.D. 1992. Forest wildfire, atmospheric CO_2 and solar irradiance periodicity (abstract). Eos, Transactions, American Geophysical Union, April 7 Supplement: 70.

Avrett, E.H. 1991. Temporal variations of near-UV, visible and infrared spectral irradiance from a theoretical viewpoint. Pp. 20-42 in SOLERS22, Proceedings of Workshop held 3-6 June, Boulder, CO, R.F. Donnelly (ed.).

Bachmann, K.T., and O.R. White. 1994. Observations of hysteresis in solar cycle variations among seven solar activity indicators. Solar Phys. 150:347-357.

Baker, D.N., J.B. Blake, D.J. Gorney, P.R. Higbie, R.W. Klebesadel, and J.H. King. 1987. Highly relativistic magnetospheric electrons: A role in coupling to the middle atmosphere. Geophys. Res. Lett. 14: 1027-1030.

Baker, D.N., R.A. Goldberg, F.A. Herrero, J.B. Blake, and L.B. Callis. 1993. Satellite and rocket studies of relativistic magnetospheric electrons, and their influence on the middle atmosphere. J. Atmos. Terrestr. Phys. 55:1619-1628.

Balachandran, N.K., and D. Rind. 1994. Effects of solar variability and the QBO on the modeled Troposphere/Stratosphere System. Part II: The Stratosphere. J. of Climate. Submitted.

Balan, N., G.J. Bailey, B. Jenkins, P.B. Rao and R.J. Moffett. 1994. Variations of ionospheric ionization and related solar fluxes during an intense solar cycle. J. Geophys. Res. 99:2243-2253.

Baldwin, M.P., and T.J. Dunkerton. 1989. Observations and statistical simulations of a proposed solar cycle/QBO/weather relationships. Geophys. Res. Lett. 16:863-866.

Baliunas, S. 1991. The past, present and future of solar magnetism: Stellar magnetic activity. Pp. 809-831 in The Sun in Time, C.P. Sonett, M.S. Giampapa, and M.S. Matthews (eds.). The University of Arizona Press, Tucson, AZ.

Baliunas, S., and R. Jastrow. 1990. Evidence for long-term brightness changes of solar-type stars. Nature 348:520-523.

Barnston, A.G., and R.E. Livezey. 1989. A closer look at the effect of the 11-year solar cycle and the quasibiennial oscillation in the Northern Hemisphere 700 mb height and extratropical North American surface temperature. J. Climate 2:1295-1313.

Barth, C.A., W.K. Tobiska, G.J. Rottman, and O.R. White. 1990. Comparison of 10.7 cm radio flux with SME solar Lyman α flux. Geophys. Res. Lett. 17:571-574.

Beck, J.G., and G.A. Chapman. 1993. A study of the contrast of sunspots from photometric images. Solar Phys. 146:49-60.

Beer, J., U. Siegenthaler, G. Bonani, R.C. Finkel, H. Oeschger, M. Suter, and W. Wolfli. 1988. Information on past solar activity and geomagnetism from ^{10}Be in the Camp Century ice core. Nature 331:675-679.

Beer, J., G.M. Raisbeck, and F. Yiou. 1991. Time variation of ^{10}Be and solar activity. Pp. 343-359 in The Sun in Time, C.P. Sonett, M.S. Giampapa, and M.S. Matthews (eds.). The University of Arizona Press, Tucson, AZ.

Berger, A., J.L. Jelice, and I. Van der Mersch. 1990. Evolutive spectral analysis of sunspot data over the past 300 years. Phil. Trans. R. Soc. Lond. A 330:529-541.

Bojkov, R., I. Bishop, W.J. Hill, G.C. Reinsel, and G.C. Tiao. 1990. A statistical trend analysis of revised Dobson total ozone data over the northern hemisphere. J. Geophys. Res. 95:9785-9807.

Bradley, R.S., and P.D. Jones. 1993. Little Ice Age summer temperature variations: their nature and relevance to recent global warming trends. The Holocene 3(4):367-376.

Brasseur, G. 1993. The response of the middle atmosphere to long term and short-term solar variability: A two dimensional model. J. Geophys. Res. 98:23079-23090.

Brueckner, G.E., K.L. Edlow, L.E. Floyd IV, J.L. Lean, and M.E. VanHoosier. 1993. The solar ultraviolet spectral irradiance monitor (SUSIM) experiment on board the Upper Atmosphere Research Satellite (UARS). J. Geophys. Res. 98:10695-10711.

Callis, L.B., R.E. Boughner, M. Natarajan, J.D. Lambeth, D.N. Baker, and J.B. Blake. 1991. Ozone depletion in the high latitude lower stratosphere: 1979-1990. J. Geophys. Res. 96:2921-2937.

Cebula, R.P., M.T. DeLand, and B.M. Schlesinger. 1992. Estimates of solar variability using the backscatter ultraviolet (SBUV) 2 Mg II index from the NOAA-9 satellite. J. Geophys. Res. 97:11613-11620.

Chandra, S. 1991. The solar UV related changes in total ozone from a solar rotation to a solar cycle. Geophys. Res. Lett. 18:837-840.

Chanin, M.L., and P. Keckhut. 1991. Influence on the middle atmosphere of the 27-day and 11-year solar cycles: radiative and/or dynamical forcing? J. Geomag. Geoelectr. Suppl. 43:647-655.

Chapman, G.A., A.D. Herzog, and J.K. Lawrence. 1986. Time-integrated energy budget of a solar activity complex. Nature 319:654-655.

Cohen, T.J., and E.I. Sweester. 1975. The spectra of the solar cycle and of data for Atlantic tropical cyclones. Nature 256:295-296.

Committee on Earth Sciences. 1989. Our Changing Planet: The FY 1990 Research Plan. The U.S. Global Change Research Program, Federal Coordinating Council on Science, Engineering and Technology.

Cook, J.W., G.E. Brueckner, and M.E. VanHoosier. 1980. Variability of the solar flux in the far ultraviolet 1175-2100 Å. J. Geophys. Res. 85:2257-2268.

Crowley, T.J., and M.K. Howard. 1990. Testing the sun-climate connection with paleoclimate data. Pp. 81-89 in Proceddings of Conference on the Climatic Impact of Solar Variability, NASA, GSFC, 24-27th April, K. Schatten, A. Arking, and R. Schiffer (eds.). NASA CP-3086.

Crowley, T.J., and R.J. North. 1991. Paleoclimatology, Oxford University Press, NY.

Crutzen, P.J. 1992. Ultraviolet on the increase. Nature 356:104-105.

Dahe, Q., E. Zeller, and G. Dreschhoff. 1991. The distribution of nitrate content in the surface snow of the Antarctic ice sheet along the route of the 1990 international trans-Antarctic expedition. J. Geophys. Res. 97:6277-6284.

Damon, P.E. 1977. Solar induced variations of energetic particles at one AU. Pp. 429-448 in The Solar Output and its Variation, O.R. White (ed.). Colorado Assoc. University Press, Boulder. CO.

Damon, P.E., and J.L. Jirikowic. 1994. Solar forcing of global climate change, Proceedings of IAU Colloquium No. 143. The Sun as a Variable Star: Solar and Stellar Irradiance Varations, Boulder, June 20-25, 1993. J. Pap, H. Hudson, and S. Solanki (eds.). Cambridge University Press.

Damon, P.E., and C.P. Sonett. 1991. Solar and terrestrial components of the ^{14}C variation spectrum. Pp. 360-388 in The Sun in Time, C.P. Sonett, M.S. Giampapa, and M.S. Matthews (eds.). University of Arizona Press, Tucson, AZ.

DeLand, M.T., and R.P. Cebula. 1993. The composite Mg II solar activity index for solar cycles 21 and 22. J. Geophys. Res. 98:12809-12823.

DeLuca, E.E., and P.A. Gilman. 1991. The Solar Dynamo. Pp. 275-303 in The Solar Interior and Atmosphere, A.N. Cox, W.C. Livingston, and M.S. Matthews (eds.). University of Arizona Press, Tucson, AZ.

Dicke, R.H. 1978. Is there a chronometer hidden deep in the Sun? Nature 276:676-680.

Donnelly, R.F. 1988. Uniformity in solar UV flux variations important to the stratosphere. Annales Geophys. 6:417-424.

Dreschhoff, G.A.M., and E.J. Zeller. 1990. Evidence of individual solar proton events in antarctic snow. Solar Phys. 127:333-346.

Eddy, J.A. 1976. The Maunder Minimum. Science 192:1189-1202.

Eddy, J.A. 1977. Historical evidence for the existence of the solar cycle. Pp. 51-71 in The Solar Output and its Variation, O.R. White (ed.). Colorado Assoc. University Press, Boulder, CO.

Eddy, J.A., P.A. Gilman, and D.E. Trotter. 1976. Solar rotation during the Maunder Minimum. Solar Phys. 46:3-14.

Evans, J.V. 1982. The Sun's influence on the Earth's atmosphere and interplanetary space. Science 216:467-474.

Feng, W., H.S. Ogawa, and D.L. Judge. 1989. The absolute solar soft X ray flux in the 20- to 100-Å region. J. Geophys. Res. 94:9125-9130.

Foukal, P. 1981. Sunspots and changes in the global output of the Sun. Pp. 391 in The Physics of Sunspots, L.E. Cram and J.H. Thomas (eds.). Sacramento Peak Observatory, New Mexico.

Foukal, P. (ed.). 1987. Solar Radiative Output Variation, Proceedings of a Workshop held Nov. 9-11, 1987, NCAR, Boulder, CO.

Foukal, P. 1994. On stellar luminosity variations and global warming. Science 264:238-239.

Foukal, P., and J. Lean. 1986. The influence of faculae on total solar irradiance and luminosity. Astrophys. J. 302:826-835.

Foukal, P., and J. Lean. 1988. Magnetic modulation of solar luminosity by photospheric activity. Astrophys. J. 328:347-357.

Foukal, P., and J. Lean. 1990. An empirical model of total solar irradiance variation between 1874 and 1988. Science 247:505-604.

Foukal, P.V., C. Hoyt, H. Kochling, and P. Miller. 1990. Cryogenic absolute radiometers as laboratory irradiance standards, remote sensing detectors, and pyroheliometers. Applied Optics 29:988-993.

Foukal, P., K. Harvey, and F. Hill. 1991. Do changes in the photospheric magnetic network cause the 11-year variation of total solar irradiance? Astrophys. J. 383:L89-L92.

Friis-Christensen, E., and K. Lassen. 1991. Length of the solar cycle: An indicator of solar activity closely associated with climate. Science 254:698-700.

Frohlich, C. 1977. Contemporary measures of the solar constant. Pp. 93-109 in The Solar Output and its Variation, O.R. White (ed.). Colorado Assoc. University Press, Boulder, CO.

Fukui, K. 1990. Anomalous increase of Lyman α flux during the solar maximum phase of cycle 21 observed by the AE-E satellite. Geophysics Laboratory Environmental Research Papers. No. 1051. GL-TR-90-0031.

Garcia, R., S. Solomon, R. Roble, and D. Rusch. 1984. A numerical response of the middle atmosphere to the 11-year solar cycle. Planet. Space Sci. 32:411-423.

Goldberg, R.A., C.H. Jackman, J.R. Barcus, and F. Soraas. 1984. Nighttime auroral energy deposition in the middle atmosphere. J. Geophys. Res. 89:5581-5596.

Gorney, D.J. 1990. Solar cycle effects on the near-earth space environment. Rev. Geophys. 28:315-336.

Gough, D.O., and J. Toomre. 1991. Seismic observations of the solar interior. Ann. Rev. Astron. Astrophys. 29:627-684.

Hansen, J., and S. Lebedeff. 1987. Global trends of measured surface air temperature. J. Geophys. Res. 92:13,345-13,372.

Hansen, J.E., and A.A. Lacis. 1990. Sun and dust versus greenhouse gases: an assessment of their relative roles in global climate change. Nature 346:713-719.

Hansen, J., A. Lacis, D. Rind, G. Russell, P. Stone, I. Fung, R. Ruedy, and J. Lerner. 1994. Climate sensitivity: analysis of feedback mechanisms. Pp. 130-163 in Climate Processes and Climate Sensitivity, J. Hansen and T. Takahashi (eds.). American Geophysical Union, Washington, DC.

Hansen, J., A. Lacis, R. Ruedy, M. Sato, and H. Wilson. 1993. How sensitive is the world's climate ? Natl. Geogr. Res. Explor. 9:142-158.

Hartmann, D.L., B.R. Barkstrom, D. Crommelynck, P. Foukal, J.E. Hansen, J. Lean, R.B. Lee III, M.R. Schoerberl, and R.C. Willson. 1993. Total solar irradiance monitoring: Report of the Atmospheres Panel to the Payload Panel, November 1993. The Earth Observer, EOS Program 5:23-27.

Harvey, J.W. 1984. Helium 10,830-Å irradiances: 1975-1983. Pp. 197-211 in Solar Irradiance Variations on Active Region Time Scales, B.J. LaBonte, G.A. Chapman, H.S. Hudson, and R.C. Willson (eds.). NASA Conf. Publ. CP-2310.

Harvey, J.W., and W.C. Livingston. 1994. Variability of the solar He I 10830 Å triplet. Pp. 59-64 in International Astronomical Union Symposium 154: Infrared Solar Physics, D.M. Rabin, J.T. Jefferies, and C. Lindsey (eds.).

Harvey, J.W., J.R. Kennedy, and J.W. Leibacher. 1987. GONG: To see inside our Sun, Sky and Telescope November: 470-476.

Hauglustaine, D.A., C. Grainer, G.P. Brasseur, and G. Megie. 1994. The importance of atmospheric chemistry in the calculation of radiative forcing on the climate system. J. Geophys. Res. 99:1173-1186.

Hays, J.D., J. Imbrie, and N.J. Shakleton. 1976. Variations in the Earth's orbit: Pacemaker of the ice ages. Science 194:1121-1132.

Heath, D.F., and B.M. Schlesinger. 1986. The Mg 280-nm doublet as a monitor of changes in solar ultraviolet irradiance. J. Geophys. Res. 91:8672-8682.

Heath, D.F., A.J. Kruger, and P.J. Crutzen. 1977. Solar proton event: influence on stratospheric ozone. Science 197:886-889.

Herman, J.R., and R.A. Goldberg. 1978. Sun, Weather and Climate, NASA SP-426.

Herman, J.R., R.D. Hudson, and G. Serafino. 1990. Analysis of the eight-year trend in ozone depletion from empirical models of Solar Backscattered Ultraviolet instrument degradation. J. Geophys. Res. 95:7403-7416.

Heroux, L., and H.E. Hinteregger. 1978. Aeronomical reference spectrum for solar UV below 2000 Å. J. Geophys. Res. 83:5305-5308.

Herrero, F.A., D.N. Baker, and R.A. Goldberg. 1991. Rocket measurements of relativistic electrons: New features in fluxes, spectra and pitch angle distributions. Geophys. Res. Lett. 18:1481-1484.

Hickey, J., R. Bradley, M. Alton, H. L. Kyle, and D. Hoyt. 1988. Total solar irradiance measurements by ERB/Nimbus-7. A review of nine years. Space Science Reviews 48:321-342.

Hinteregger, H.E., K. Fukui, and B.G. Gilson. 1981. Observational, reference and model data on solar EUV, from measurements on AE-E. Geophys. Res. Lett. 8:1147-1150.

Hoegy, W.R., W.D. Pesnell, T.N. Woods, and G.J. Rottman. 1993. How active was solar cycle 22? Geophys. Res. Lett. 20:1335-1338.

Holdsworth, G., H.R. Krouse, M. Nosal, M.J. Spenser, and P.A. Mayewski. 1989. Analysis of a 290-year net accumulation time series from Mt. Logan, Yukon. Pp. 71 in Snow and Glacier Variations, IAHS Publ. # 183.

Hood, L.L. 1987. Solar ultraviolet radiation induced variations in the stratosphere and mesosphere. J. Geophys. Res. 92:876-888.

Hood, L., and A.R. Douglass. 1988. Stratospheric response to solar ultraviolet variations: Comparisons with photochemical models. J. Geophys. Res. 93:3905-3911.

Hood, L.L., and J.P. McCormack. 1992. Components of interannual ozone change based on Nimbus 7 TOMS data. Geophys. Res. Lett. 19:2309-2312.

Hood, L.L., J.L. Jirikowic, and J.P. McCormack. 1993. Quasi-decadal variability of the stratosphere: Influence of long term solar ultraviolet variations. J. Atmos. Sci. 50:3941-3958.

Hoyt, D.V. 1979. The Smithsonian Astrophysical Observatory solar constant program. Rev. Geophys. Space Phys. 17:427-458.

Hoyt, D.V., and K.H. Schatten. 1993. A discussion of plausible solar irradiance variations 1700-1992. J. Geophys. Res. 98:18895-18906.

Hoyt, D.V., H.L. Kyle, J.R. Hickey, and R.H. Maschhoff. 1992. The Nimbus 7 solar total irradiance: A new algorithm for its derivation. J. Geophys. Res. 97:51-63.

Hoyt, D.V., K.H. Schatten, and E. Nesmes-Ribes. 1994. The one hundredth year of Rudolf Wolf's death: Do we have the correct reconstruction of solar activity? Geophys. Res. Lett. In press.

Huang, T.Y., and G.P. Brasseur. 1993. Effect of long term solar variability in a two-dimensional interactive model of the middle atmosphere. J. Geophys. Res. 98:20413-20427.

Hudson, H.S. 1987. Solar variability and oscillations. Rev. Geophys. 25:651-662.

Intergovernmental Panel on Climate Change, Climate Change 1992. The Supplementary Report to the IPCC Scientific Assessment. J.T. Houghton, B.A. Callander, and S.K. Varney (eds.). Cambridge University Press.

Jackman, C.H. 1991. Effects of energetic particles on minor constituents of the middle atmosphere. J. Geomag. Geoelectr. Suppl. 43:637-646.

Jackman, C.H., and R.D. McPeters. 1987. Solar proton events as tests for the fidelity of middle atmosphere models. Physica Scripta T18:309-316.

Jackman, C.H., A.R. Douglass, R.B. Rood, R.D. McPeters, and P.E. Meade. 1990. Effect of solar proton events on the middle atmosphere during the past two solar sycles as computed using a two-dimensional model. J. Geophys. Res. 95:7417-7428.

James, I.N., and P.M. James. 1989. Ultra-low-frequency variability in a simple atmospheric circulation model. Nature 342:53-55.

Johnsen, S.J., W. Dansgaard, H.B. Clausen, and C.C. Langway. 1970. Climatic oscillations 1200-2000 A.D. Nature 227:482-483.

Joselyn, J.A. 1990. Case study of the great geomagnetic storm of 13 March 1989. Pp. 745-762 in Astrodynamics 1989. 71. Advances in the Astronautical Sciences, C.L. Thorton, R.J. Proulx, J.E. Prussing, and F.R. Hoots (eds.). American Astronautical Society, Univelt Publish.

Joselyn, J.A., and E.C. Wipple. 1990. Effects of the space environment on space science. American Scientist 78:126-133.

Justus, C.G., and B.B. Murphey. 1994. Temporal trends in surface irradiance at ultraviolet wavelengths. J. Geophys. Res. 99:1389-1394.

Kelly, P.M. 1977. Solar influences on north Atlantic mean sea level pressure. Nature 269:320-322.

Kelly, P.M., and T.M.L. Wigley. 1992. Solar cycle length, greenhouse forcing and global climate. Nature 360:328-330.

Kodera, K. 1991. The solar and equatorial QBO influences on the stratospheric circulation during the early northern-hemisphere winter. Geophys. Res. Lett. 18:1023-1026.

Kodera, K., and K. Yamazaki. 1990. Long term variation of upper stratospheric circulation in the Northern Hemisphere in December. J. Met. Soc. of Japan 68:101-105.

Kremliovsky, M.N. 1994. Can we understand time scales of solar activity? Solar Phys. 151:351-370.

Kuhn, J.R., and K.G. Libbrecht. 1991. Nonfacular solar luminosity variations. Astrophys. J. 381:L35-L37.

Kuhn, J.R., K.G. Libbrecht, and R.H. Dicke. 1988. The surface temperature of the sun and changes in the solar constant. Science 242:908-911.

Kurucz, R.L. 1991. The solar spectrum. Pp. 663-669 in The Solar Interior and Atmosphere, A.N. Cox, W.C. Livingston, and M.S. Matthews (eds.). University of Arizona Press, Tucson, AZ.

Labitzke, K., and H. van Loon. 1990. Associations between the 11-year solar cycle, the quasibiennial oscillation and the atmosphere: a summary of recent work. Phil. Trans. Royal Society of London A 330:577-589.

Labitzke, K., and H. van Loon. 1993. Some recent studies of probable connections between solar and atmospheric variability. Ann. Geophysicae 11:1084-1094.

Lacis, A.A., and B.E. Carlson. 1992. Global warming: Keeping the Sun in proportion. Nature 360:297.

Lacis, A.A., D.J. Wuebbles, and J.A. Logan. 1990. Radiative forcing of climate by changes in the vertical distribution of ozone. J. Geophys. Res. 95:9971-9981.

Lean, J. 1987. Solar ultraviolet irradiance variations: A review. J. Geophys. Res. 92:839-868.

Lean, J. 1988. Solar EUV Irradiances and Indices. Adv. Space Res. 8:(5)263-(5)292.

Lean, J. 1989. Contribution of ultraviolet irradiance variations to changes in the Sun's total irradiance. Science 244:197-200.

Lean, J. 1990. A comparison of models of the Sun's extreme ultraviolet irradiance variations. J. Geophys. Res. 95:11933-11944.

Lean, J. 1991. Variations in the Sun's radiative output. Reviews of Geophys. 29:505-535.

Lean, J., and A. Skumanich. 1983. Variability of the Lyman α flux with solar activity. J. Geophys. Res. 88:5751-5563.

Lean, J.L., O.R. White, W.C. Livingston, D.F. Heath, R.F. Donnelly, and A. Skumanich. 1982. A three-component model of the variability of the solar ultraviolet flux: 145-200 nm. J. Geophys. Res. 87:10307-10317.

Lean, J., A. Skumanich, and O.R. White. 1992a. Estimating the Sun's radiative output during the Maunder minimum. Geophys. Res. Lett. 19:1591-1594.

Lean, J., M. VanHoosier, G. Brueckner, D. Prinz, L. Floyd, and K. Edlow. 1992b. SUSIM/UARS observations of the 120 to 300 nm flux variations during the maximum of the solar cycle: Inferences for the 11-year cycle. Geophys. Res. Lett. 19:2203-2206.

Lee III, R.B. 1990. Long-term solar irradiance variability: 1984-1989 Observations. Pp. 301-308 in Proceedings of Conference on the Climatic Impact of Solar Variability, NASA, GSFC, 24-27th April, K. Schatten, A. Arking, and R. Schiffer (eds.). NASA CP-3086.

Lee III, R.B., M.A. Gibson, R.S. Wilson, and S. Thomas. 1994. Long term solar irradiance variability during sunspot cycle 22. J. Geophys. Res. Submitted.

Leibacher, J.W., R.W. Noyes, J. Toomre, and R.K. Ulrich. 1985. Helioseismology. Scientific American 253:48-57.

Livingston, W.C., L. Wallace, and O.R. White. 1988. Spectrum line intensity as a surrogate for solar irradiance variations. Science 240:1765-1767.

Lockwood, G.W., and B.A. Skiff. 1990. Some insights on solar variability from precision stellar astronomical photometry. Pp. 8-15 in Proceedings of Conference on the Climatic Impact of Solar Variability, NASA, GSFC, 24-27th April, K. Schatten, A. Arking, and R. Schiffer (eds.). NASA CP-3086.

Lockwood, G.W., B.A. Skiff, S.L. Baliunas, and R.R. Radick. 1992. Long-term solar brightness changes estimated from a survey of Sun-like stars. Nature 360:653-655.

London, J., and G.J. Rottman. 1990. Wavelength dependence of solar rotation and solar cycle UV irradiance variations. Pp. 323-327 in Proceedings of Conference on the Climatic Impact of Solar Variability. NASA, GSFC, 24-27th April, K. Schatten, A. Arking, and R. Schiffer (eds.). NASA CP-3086.

London, J., G.J. Rottman, T.N. Woods, and F. Wu. 1993. Time variations of solar UV irradiance as measured by the SOLSTICE (UARS) instrument. Geophys. Res. Lett. 20:1315-1318.

Lopate, C., and J.A. Simpson. 1991. The physics of cosmic ray modulation: Heliospheric propagation during the 1987 minimum. J. Geophys. Res. 96:15877-15898.

Luther, M.R., R.B. Lee III, B.R. Barkstrom, J.E. Cooper, R.D. Cess, and C.H. Duncan. 1986. Solar calibration results from two earth radiation budget nonscanner instruments. App. Optics 25:540-545.

Madronich, S. 1992. Implications of recent total atmosphere ozone measurements for biologically active ultraviolet radiation reaching the Earth's surface. Geophys. Res. Lett. 19:37-40.

Madronich, S., and F.R. de Gruijl. 1993. Skin cancer and UV radiation. Nature 366:23.

McHargue, L.R., and P.E. Damon. 1991. The global Beryllium 10 cycle. Rev. Geophys. 29:141-158.

Meier, R.R. 1991. Ultraviolet spectroscopy and remote sensing of the upper atmosphere. Space Science Reviews 58:1-185.

Mitchell, W.E., and W.C. Livingston. 1991. Line-blanketing in the irradiance spectrum of the sun from maximum to minimum of the solar cycle. Astrophys. J. 372:336-348.

Mitchell, J.M., C.W. Stockton, and D.M. Meko. 1979. Evidence of a 22-year rhythm of drought in the western United States related to the Hale solar cycle since the 17th century. Pp. 124-144 in Solar Terrestrial Influences on Weather and Climate, B.M. McCormac and T.A. Seliga (eds.). D. Reidel, Hingham, MA.

Mohanakumar, K. 1989. Influence of solar activity on middle atmosphere associated with phase of equatorial QBO. MAP Handbook, 29:39-42.

Moore, W.S. 1982. Chapter 18 in Uranium Series Disequilibrium: Application in Environmental Problems. M. Ivanovich and R.S. Harmon (eds.). Clarendon, Oxford, U.K.

Morfill, G.E., H. Scheingraber, W. Voges, and C.P. Sonnet. 1991. Sunspot number variations: Stochastic or chaotic. Pp. 30-58 in The Sun in Time, C.P. Sonett, M.S. Giampapa, and M.S. Matthews (eds.). University of Arizona Press, Tucson, AZ.

Mundt, M.D., W.B. Maguire II, and R.P. Chase. 1991. Chaos in the sunspot cycle: Analysis and prediction. J. Geophys. Res. 96:1705-1716.

National Academy of Sciences. 1982. Studies in Geophysics: Solar Variability, Weather, and Climate. National Academy Press, Washington, DC.

National Academy of Sciences. 1986. Studies in Geophysics: The Earth's Electrical Environment. National Academy Press, Washington, DC.

National Academy of Sciences. 1988. Long-Term Solar-Terrestrial Observations. National Academy Press, Washington, DC.

National Academy of Sciences. 1991. Space Studies Board. Assessment of Programs in Solar and Space Physics. National Academy Press, Washington, DC.

Nesme-Ribes, E., E.N. Ferreira, R. Sadourny, H. Le Truet, and Z.X. Li. 1993. Solar dynamics and its impact on solar irradiance and the terrestrial climate. J. Geophys. Res. 98:18923-18935.

Newell, N.E., R.E. Newell, J. Hsuing, and W. Zhongxiang. 1989. Global marine temperature variation and the solar magnetic cycle. Geophys. Res. Lett. 16:311-314.

Newkirk, G. 1983. Variations in solar luminosity. Annual Review of Astronomy and Astrophysics 21:429-467.

Ogawa, H.S., L.R. Canfield, D. McMullin, and D.L. Judge. 1990. Sounding rocket measurements of the absolute solar EUV flux utilizing a silicon photodiode. J. Geophys. Res. 95:4291-4295.

Otaola, J.A., and G. Zenteno. 1983. On the existence of long term periodicities in solar activity. Solar Phys. 89:209-213.

Petersen, C., M. Bruner, L. Acton, and Y. Ogawara. 1993. Yohkoh and the mysterious solar flares. Sky and Telescope, September: 20-25.

Phillipps, P., and I.M. Held. 1994. The response to orbital perturbations in an atmospheric model coupled to a slab ocean. J. Climate 7:767-782.

Price, C. and D. Rind. 1994. Possible implications of global climate change on global lightning distributions and frequencies. J. Geophys. Res. 99:10823-10831.

Radick, R.R., G.W. Lockwood, and S.L. Baliunas. 1990. Stellar activity and brightness variations: A glimpse at the sun's history. Science 247:39-44.

Randel, W.J., and J.B. Cobb. 1994. Coherent variations of monthly mean total ozone and lower stratospheric temperature. J. Geophys. Res. 99:5433-5447.

Reid, G.C. 1974. Polar cap absorption - Observations and theory. Fundamentals of Cosmic Physics 1:167-202.

Reid, G.C. 1991. Solar total irradiance variations and the global sea surface temperature record. J. Geophys. Res. 96:2835-2844.

Reid, G.C., S. Solomon, and R.R. Garcia. 1991. Response of the middle atmosphere to the solar proton events of August-December 1989. Geophys. Res. Lett. 18:1019-1022.

Reinsel, G.C., G.C. Tiao, S.K. Ahn, M. Pugh, S. Basu, J.J. DeLuisi, C.L. Mateer, A.J. Miller, P.S. Connell, and D.J. Wuebbles. 1988. An analysis of the 7-year record of SBUV satellite ozone data: Global profile features and trends in total ozone. J. Geophys. Res. 93:1689-1703.

Reinsel, G.C., G.C. Tiao, D.J. Wuebbles, J.B. Kerr, A.J. Miller, R.M. Nagatani, L. Bishop, and L.H. Ying. 1994a. Seasonal trend analysis of published ground-based and TOMS total ozone data through 1991. J. Geophys. Res. 99:5449-5464.

Reinsel, G.C., W.K. Tam, and L.H. Ying. 1994b. Comparison of trend analyses for Umkehr data using new and previous inversion algorithms. Geophys. Res. Lett. 21:1007-1010.

Ribes, E., P. Merlin, J.C. Ribes, and R. Barthalot. 1989. Absolute periodicities in the solar diameter derived from historical and modern data. Ann. Geophys. 7:321-330.

Richards, P.G., and D.G. Torr. 1988. Ratios of photoelectron to EUV ionization rates for aeronomic studies. J. Geophys. Res. 93:4060-4066.

Rind, D., and N.K. Balachandran. 1994. Effects of solar variability and the QBO on the modeled Troposphere/Stratosphere System. Part II: The Troposphere. J. of Climate. Submitted.

Rind, D., and J. Overpeck. 1993. Hypothesized causes of decade-to-century climate variability: Climate model results. Quat. Sci. Rev. 12:357-374.

Rind, D., D. Peteet, and G. Kukla. 1989. Can Milankovitch orbital variations initiate the growth of ice sheets in a general circulation model? J. Geophys. Res. 94:12851-12871.

Rind, D., R. Suozzo, N.K. Balachandran, and M.J. Prather. 1990. Climate change and the middle atmosphere. Part I: The doubled CO_2 climate. J. Atmos. Sci. 47:475-494.

Rishbeth, H. 1990. A greenhouse effect in the ionosphere? Planet. Space Sci. 38:945-948.

Roble, R.G., and R.E. Dickinson. 1989. How will changes in carbon dioxide and methane modify the mean structure of the mesosphere and thermosphere? Geophys. Res. Lett. 16:1441-1444.

Rottman, G.J. 1988. Observations of solar UV and EUV variability. Adv. Space Res. 8:(7)53-(7)66.

Rottman, G.J., T.N. Woods, and T.P. Sparn. 1993. Solar-Stellar Irradiance Comparison Experiment 1. Instrument Design and Operation. J. Geophys. Res. 98:10667-10677.

Royal Society. 1990. The Earth's Climate and Variability of the Sun over Recent Millennia. Phil. Trans. Royal Society of London A, Volume 330, Cambridge University Press.

Salby, M.L., and D.J. Shea. 1991. Correlations between solar activity and the atmosphere: An unphysical explanation. J. Geophys. Res. 96:22579-22595.

Schatten, K.H. 1988. A model for solar constant secular changes. Geophys. Res. Lett. 15:121-124.

Schatten, K.H. 1993. Heliographic latitude dependence of the Sun's irradiance. J. Geophys. Res. 98:18907-18910.

Schatten, K.H., and W.D. Pesnell. 1993. An early solar dynamo prediction: Cycle 23 $\sim$ cycle 22. Geophys. Res. Lett. 20:2275-2278.

Schlesinger, M.E., and N. Ramankutty. 1992. Implications for global warming of intercycle solar irradiance variations. Nature 360:330-333.

Schmidtke, G., T.N. Woods, J. Worden, G.J. Rottman, H. Doll, C. Wita, and S.C. Solomon. 1992. Solar EUV irradiance from the San Marco ASSI: A reference spectrum. Geophys. Res. Lett. 19:2175-2178.

Schwarzkopf, M.D., and V. Ramaswamy. 1993. Radiative forcing due to ozone in the 1980's: Dependence on altitude of ozone change. Geophys. Res. Lett. 20:205-208.

Sharber, J.R., R.A. Frahm, J.D. Winningham, J.C. Biard, D. Lummerzheim, M.H. Rees, D.L. Chenette, E.E. Gaines, R.W. Nightingale, and W.L. Imhof. 1993. Observations of the UARS Particle Environment Monitor and computation of ionization rates in the middle and upper atmosphere during a geomagnetic storm. Geophys. Res. Lett. 20:1319-1322.

Siskind, D.E. 1994. On the radiative coupling between mesospheric and thermospheric nitric oxide. J. Geophys. Res. In press.

Siskind, D.E., C.A. Barth, and D.D. Cleary. 1990. The possible effect of solar soft X-rays on thermospheric nitric oxide. J. Geophys. Res. 95:4311-4317.

Smith, P.L., J.L. Lean, A.B. Christensen, K.L. Harvey, D.L. Judge, R.L. Moore, M.R. Torr, and T.N. Woods. 1993. SOURCE: The Solar Ultraviolet Radiation and Correlative Emissions Mission. Metrologia 30:275-277.

Sofia, S., and P. Fox. 1994. Solar Variability and Climate. Climatic Change 30:1-9.

Sofia, S., L. Oster, and K. Schatten. 1982. Solar irradiance modulation by active regions during 1980. Solar Phys. 80:87-98.

Solomon, S., and R. Garcia. 1984. Transport of thermospheric NO to the upper stratosphere. Planet. Space Sci. 32:399-409.

Stephenson, J.A., and M.W.J. Scourfield. 1991. Importance of energetic solar protons in ozone depletion. Nature 352:137-139.

Stolarski, R.S., P. Bloomfield, R.D. McPeters, and J.R. Herman. 1991. Total ozone trends deduced from Nimbus 7 TOMS data. Geophys. Res. Lett. 18:1015-1018.

Stuiver, M., and T.F. Braziunas. 1993. Sun, ocean, climate and atmosphere $^{14}CO_2$: an evaluation of causal and spectral relationships. The Holocene 3(4):289-305.

Stuiver, M., and P.J. Reimer. 1993. Extended ^{14}C data base and revised CALIB 3.0 ^{14}C age calibration program. Radiocarbon 35:215-230.

Suess, H.E., and T.W. Linick. 1990. The ^{14}C record in bristlecone pine of the past 8000 years based on the dendrochronology of the late C.W. Ferguson, in The Earth's Climate and Variability of the Sun over Recent Millennia: Geophysical, Astronomical and Archaeological Aspects. Phil. Trans. R. Soc. Lond. A 330:403-412.

Thomas, R.J., C.A. Barth, G.J. Rottman, D.W. Rusch, G.H. Mount, G.M. Lawrence, R.W. Sanders, G.E. Thomas, and L.E. Clemens. 1983. Mesospheric ozone depletion during the solar proton event of July 13, 1982, 1, Measurements. Geophys. Res. Lett. 10:253-255.

Thompson, L.G., E. Mosley-Thompson, M.E. Davis, T. Yao, P.N. Lin, and J. Dai. 1993. New evidence for recent warming from high resolution Chinese and Peruvian Ice Cores (abstract). Eos, Transactions, American Geophysical Union, 1993 Spring Meeting Supplement: 90.

Thompson, R.J. 1993. A technique for predicting the amplitude of the solar cycle. Solar Phys. 148:383-388.

Thorne, R.M. 1980. The importance of energetic particle precipitation on the chemical composition of the middle atmosphere. Pure Appl. Geophys. 118:128-151.

Tinsley, B.A., and G.W. Deen. 1991. Apparent tropospheric response to MeV-GeV particle flux variations: A connection via electrofreezing of supercooled water in high-level clouds. J. Geophys. Res. 96:22283-22296.

Tobiska, W.K. 1991. A revised solar extreme ultraviolet flux model. J. Atmos. Terr. Phys. 53:1005-1018.

van Loon, H., and K. Labitzke. 1988. Association between the 11-year solar cycle, the QBO and the atmosphere, Part II: Surface and 700 mb in the Northern Hemisphere in Winter. J. Climate 1:905-920.

Vidal-Madjar, A. 1975. Evolution of the solar Lyman α flux during four consecutive years. Solar Phys. 40:69-86.

Vidal-Madjar, A., and B. Phissamay. 1980. The solar Lyman α flux near solar minimum. Solar Phys. 66:259-271.

Weeks, L.H., R.S. Cuikay, and J.R. Corbin. 1972. Ozone measurements in the mesosphere during the solar proton event of November 2, 1969. J. Atmos. Sci. 29:1138-1142.

White, O.R., and W.C. Livingston. 1981. Solar luminosity variation, III, Calcium K variation from solar minimum to solar maximum in cycle 21. Astrophys. J. 249:798-816.

White, O.R., G.J. Rottman, and W.C. Livingston. 1990. Estimation of the solar Lyman α flux from ground based measurements of the Ca II K line. Geophys. Res. Lett. 17:575-578.

White, O.R., A. Skumanich, J. Lean, W.C. Livingston, and S.L. Keil. 1992. The sun in a non-cycling state. Publications of the Astronomical Society of the Pacific 104:1139-1143.

White, O.R, G.J. Rottman, T.N. Woods, B.G. Knapp, S.L. Keil, W.C. Livingston, K.F. Tapping, R.F. Donnelly, and L.C. Puga. 1994. Change in the radiative output of the Sun in 1992 and its effect in the thermosphere. J. Geophys. Res. 99:369-372.

Wigley, T.M.L., and P.M. Kelly. 1990. Holocene climatic change, ^{14}C wiggles and variations in solar irradiance, in The Earth's Climate and Variability of the Sun over Recent Millennia: Geophysical, Astronomical and Archaeological Aspects. Phil. Trans. Royal Society of London A. 330:547-560.

Wigley, T.M.L., and S.C.B. Raper. 1990. Climatic change due to solar irradiance changes. Geophys. Res. Lett. 17:2169-2172.

Wilcox, J., L. Svalgaard, and P. Scherrer. 1976. On the reality of a sun-weather effect. J. Atmos. Sci. 33:1113-1116.

Willson, R.C. 1984. Measurements of solar total irradiance and its variability. Space Science Rev. 38:203-242.

Willson, R.C., and H.S. Hudson. 1991. A solar cycle of measured and modeled total irradiance. Nature 351:42-44.

Willson, R.C., S. Gulkis, M. Janssen, H.S. Hudson, and G.A. Chapman. 1981. Observations of solar irradiance variability. Science 211:700-702.

Winningham, J.D., D.T. Decker, J.U. Kozyra, J.R. Jasperse, and A.F. Nagy. 1989. Energetic (> 60 eV) atmospheric photoelectrons. J. Geophys. Res. 94:15335-15348.

Winograd, I.J., B.J. Szabo, T.B. Coplen, and A.C. Riggs. 1988. A 250,000 year climatic record from Great Basin vein calcite: Implications for Milankovitch theory. Science 242:1275-1280.

Winograd, I.J., T.B. Coplen, J.M. Landwehr, A.C. Riggs, K.R. Ludwig, B.J. Szabo, P.T. Kolesar, and K.M. Revesz. 1992. Continuous 500,000-year climate record from vein calcite in Devils Hole, Nevada. Science 258:255-260.

Withbroe, G. 1989. Report of the Working Group on Predictions of Solar Activity and the Atmosphere Response: Results of the November 8-9, 1989 meeting of the NASA Working Group.

WMO Global Ozone Research and Monitoring Project. 1988. Report No. 18. Report of the International Ozone Trends Panel.

Woods, T.N., and G.J. Rottman. 1990. Solar EUV irradiances derived from a sounding rocket experiment on 10 November 1988. J. Geophys. Res. 95:6227-6236.

Woods, T.N., G.J. Rottman, and G. Ucker. 1993. Solar-Stellar Irradiance Comparison Experiment 2, Instrument Calibration. J. Geophys. Res. 98:10679-10694.

Woods, T.N., G.J. Rottman, R. Roble, O.R. White, S.C. Solomon, G. Lawrence, J. Lean, and K. Tobiska. 1994. The TIMED Solar EUV Experiment. SPIE Paper 2266-57.

Wuebbles, D.J., D.E. Kinnison, K.E. Grant, and J. Lean. 1991. The effect of solar flux variations and trace gas emissions on recent trends in stratospheric ozone and temperature. J. Geomag. Geoelectr. Supp. 43:709-718.

Acronyms

ACRIM	Active Cavity Radiometer Irradiance Monitor
AE-E	Atmospheric Explorer -E
ARRCC	Analysis of Rapid and Recent Climatic Change
AU	Astronomical Unit
BGC	Board on Global Change (of NRC)
BP	Before Present
CEDAR	Coupling, Energetics and Dynamics of Atmospheric Regions
CFC	Chlorofluorocarbon
CSTR	Committee on Solar Terrestrial Research (of NRC)
DoD	Department of Defense
DMSP	Defense Meteorological Satellite Program
EOS	Earth Observing System
EPA	Environmental Protection Agency
ERB	Earth Radiation Budget
ERBE	Earth Radiation Budget Experiment
ERBS	Earth Radiation Budget Satellite
ESA	European Space Agency
ESSC	Electrically Self-Calibrating Cavity
EUV	Extreme Ultraviolet
EW	Equivalent Width
GCM	General Circulation Model
GCR	Galactic Cosmic Rays

GEM	Geospace Environment Modeling
GeV	Billion Electon Volts
GGS	Global Geospace Study
GOES	Geostationary Operational Environmental Satellite
GONG	Global Oscillation Network Group
ICSU	International Council of Scientific Unions
IR	Infrared
ISAS	Institute of Space and Astronautical Science (Japanese Space Agency)
ISTP	International Solar Terrestrial Program
MAHRSI	Middle Atmosphere High Resolution Spectrographic Instrument
MeV	Million Electron Volts
MSV	Mechanisms of Solar Variability
NAS	National Academy of Sciences
NDSC	Network for Detection of Stratospheric Change
NIST	National Institute of Standards and Technology
NOAA	National Oceanic and Atmospheric Administration
NRC	National Research Council
NSF	National Science Foundation
NSO	National Solar Observatory
PEM	Particle Environment Monitor
QBO	Quasibiennial Oscillation
RAIDS	Remote Atmospheric and Ionospheric Detection System
R_E	Radius of the Earth
REP	Relativistic Electron Precipitation
RISE	Radiative Inputs of the Sun to Earth
SAGE	Stratospheric Aerosol and Gas Experiment
SAMPEX	Solar, Anomalous and Magnetospheric Particles Explorer
SBUV	Solar Backscatter Ultraviolet
SCOSTEP	Scientific Council on Solar Terrestrial Physics
SEE	Spectrometer for Energetic Electrons
SERDP	Strategic Environmental Research and Development Program
SME	Solar Mesosphere Explorer
SMM	Solar Maximum Mission

SOHO	Solar and Heliospheric Observatory
SOLSTICE	Solar Stellar Irradiance Comparison Experiment
SOURCE	Solar Ultraviolet Radiation and Correlative Emissions
SPARC	Stratospheric Processes and their Relation to Climate
STEP	Solar-Terrestrial Energy Program
SUSIM	Solar Ultraviolet Spectral Irradiance Monitor
TIMED	Thermosphere Ionosphere Mesosphere Energetics and Dynamics
TOMS	Total Ozone Mapping Spectrometer
TTO	Ten to Twelve Year Oscillation
UARS	Upper Atmosphere Research Satellite
USGCRP	United States Global Change Research Program
UV	Ultraviolet
WCRP	World Climate Research Program
WMO	World Meteorological Organization